Design Science and Innovation

Series Editor

Amaresh Chakrabarti, Centre for Product Design and Manufacturing, Indian Institute of Science, Bangalore, India

The book series is intended to provide a platform for disseminating knowledge in all areas of design science and innovation, and is intended for all stakeholders in design and innovation, e.g. educators, researchers, practitioners, policy makers and students of design and innovation. With leading international experts as members of its editorial board, the series aims to disseminate knowledge that combines academic rigour and practical relevance in this area of crucial importance to the society.

More information about this series at 7 http://www.springer.com/series/15399

Sandra G. L. Persiani

Design of Autoreaction

A Framework for Kinetic Reaction at Zero Energy

Springer

Sandra G. L. Persiani
Department of Architecture
Technical University of Munich
Munich, Bayern, Germany

ISSN 2509-5986 ISSN 2509-5994 (electronic)
Design Science and Innovation
ISBN 978-981-15-6177-1 ISBN 978-981-15-6178-8 (eBook)
https://doi.org/10.1007/978-981-15-6178-8

This Springer imprint is published by the registered company Springer Nature Singapore Pte Ltd.
The registered company address is: 152 Beach Road, #21-01/04 Gateway East, Singapore 189721, Singapore

Preface

Simplicity is the ultimate sophistication.
—Leonardo da Vinci

As a species that recreates its own artificial environment, we innately strive to optimize our reality in one way or another—regardless of our field of profession or interest. To find solutions to our problems, we choose between endless possible combinations by intuitively making choices, relying on our "mental archive" of observed answers: our knowledge and our experiences. As our brains tend to save on operating energy, we consciously or unconsciously tend to fall back on old pre-tested patterns. To come up with new solutions, think creatively and ultimately innovate, combinations of previously unconnected mental representations that form new patterns of neural activity are necessary.

If a good part of innovation is about having the right tools and the right background knowledge to achieve great solutions nobody else would think of, this work aims to collect the multidisciplinary state-of-the-art knowledge at the base of a better and conscious design of autoreactive motion.

The idea to integrate moving features and a liveliness in our artefacts and tools has become an increasingly diffused concept in many fields. Motion at zero energy input—autoreaction—is however a new approach to kinetic design. It aims to push designers to come up with smarter solutions that make the most out of the available knowledge rather than delegating the kinetic execution of their designs to technology alone. Designing autoreaction requires—ideally—no more energy than the intellectual effort to identify the right concept and an efficient integration of design parameters.

The topic has been approached from a generalist's point of view, prioritizing the description of multiple possibilities involved in the design over the in-depth understanding of single parameters. The book addresses an audience of professionals, whether specialists or not, who are interested in transferring interdisciplinary concepts and ideas to their own field. It highlights the features and parameters involved in the design of autoreactive motion and can as such be read sequentially or used as a catalogue of features.

As in any interdisciplinary study, writing on a very broad topic and drawing parallels between parameters that—although related—can have very different logic and complexity has been a tiresome work raising many doubts along the way. Mainly, not being a trained mechanical engineer, energy engineer or material scientist has had its advantages and disadvantages. This analysis is knowingly a partial point of view of very multifaceted and complex realities, interpreted through an architect's vision of what autoreactivity can involve. Hence, many contents will appear oversimplified to experts in each field. Nevertheless, I strongly believe that overall this book can contribute with many suggestions that can be taken from the other topics discussed, being of inspiration to many neighbouring fields of study. I hope that this work will be useful to my fellow colleagues and to anyone else who wishes to make our artificial environment more efficient and adaptative.

The book is partially based on my Ph.D. thesis in environmental design, developed at the University of Rome La Sapienza in collaboration with the Technical University of Munich. The thesis entitled "Autoreactive Architectural Components, Theories and Schemes for the Implementation of Kinetic Reaction with Zero Energy" was defended in Rome on 22 July 2016.

All this would not have been possible without the direct and indirect support of a great number of people and institutions:

Prof. Thomas Auer and the rest of the team at the TUM Chair of Building Technology and Climate-Responsive Design for offering me the resources and the support to carry out this last year of work;

Prof. Alessandra Battisti, my former research supervisor, for enthusiastically encouraging and endorsing me during fourteen years of professional growth;

Bilge Kobas for her graphic contribution to the case-study images and her unwavering support to pursue unconventional ideas; Sebastian Koth for his valuable technical support and smart comments; and Armin Aschenbrenner for tirelessly proofreading many pages of cryptical text.

A special thanks to Luigi for never letting me down and for his profound belief in my abilities; and to my boys for reminding me to take regular breaks.

This work was supported by the TUM University Foundation Fellowship (TUFF) and the Architecture Research Incubator (ARI), Department of Architecture, Technical University of Munich (TUM).

Sandra G. L. Persiani
Munich, Germany

Introduction

Abstract
The environments and realities we live in throughout our lives are in a constant state of change, whether this happens within our range of perception or not. This chapter introduces the importance of understanding and taking into account change in the conception and shaping of the man-made environment, whether it is defined as transformation or as motion.

Stability and durability are commonly thought of as positive and sought-after features, both in the artificial world built around us as well as in nature, where we wish for a predictable and cyclical change that corresponds to stability on a broader scale. In most cases, stability can be described as a situation of equilibrium, where forces pushing in different directions balance each other out. However, if some forces are altered and no compensating forces arise, change will happen. In other words, stability and change are strongly interdependent (Schein 2002). Stability can in that aspect rather be described as a measure of how well a system can accommodate evolution or development, without requiring changes to the architecture of the system itself. Therefore, to achieve one process we need to learn to manage the other.

Change is a Necessity

Change as a Selective Value

The world as we see it today has been shaped during thousands of millions of years of automatically generated selective processes under the pressure of constant change.

Automatically generated non-random forms of order can be observed in abiotic components in nature as well as in biotic evolution. Cumulative selection refers to a non-random outcome that is generated by individual (and mostly random) actions collected over time, where the result of one action is the start for the next one. Cumulative selection can be seen in the constant action of the wind and the waves automatically arranging the pebbles on a beach according to their physical proprieties, as well as in the evolutionary process sorting entities and species over many generations, rapidly cutting incongruous combinations out from the probabilistic variation process (Dawkins 1996).

Biological organisms in particular have developed unique capacities and reactivity to change in their environment. While the cumulative selection is certainly the method applied to achieve such embedded reactivity, the tools to achieve the change are the creativity of genetic mutation and its destructive counterpart, internal and external selection. Through mutation, an organism produces change by making a small random "transcription error" during the duplication of a parental gene. This potentially "defective clone", which has the capacity to alter the developmental characters, the morphology, the colour, etc., is then tested against internal and external factors to verify how to fit the new combination is. If we think of organisms as machines, fitter genotypes correspond to machines whose parts integrate better with each other (Arthur 1997).

Change as a Human Physiological and Psychological Necessity

The genes of the contemporary humans have evolved under selective pressures and environmental conditions of much harsher magnitude than those that we expose ourselves to today. Only about fifteen generations ago, we would spend most of our time outdoors in steppes, forests and mountains, while our shelters would provide little comfort, if only providing security during the hours of darkness (Baker 2000).

Although few people would choose to return to such harsh living conditions, it would seem that—to some extent—we have evolved to associate selective drives such as hunger, sex and physical effort with pleasant experiences. In fact, the total deprivation of physical stimuli is considered to be a form of torture, and many activities carried out voluntarily to bring satisfaction largely involve the exposure to strong physical stimuli of natural origin (Baker and Standeven 1996). People from countries with cold weather conditions flock to warm weather travel destinations at the hottest time of the year, to willingly expose themselves to what can be considered quite severe conditions of discomfort (Baker 2000). Stress due to adaptive challenges is thought to derive not from the extremity of the condition themselves, but from the stimulus exceeding the objective or subjective capacity of the organism to adapt to the conditions (Baker and Standeven 1996).

In other words, responsiveness to change in the environment by complex adaptive activity is something that is natural and necessary to us: people like change and search for adaptive opportunities.

Change and the Man-Made Environment

The aptness of purposefully and systematically altering our surroundings has been one of the most typical traits of our species. Hardly any aspect in the world surrounding us today has not been shaped my man, from the things we create—tools, constructions, extensions of our bodies, boosting of our physical capacities—to the ones we alter to fit our needs—selective breeding, genetic modification, synthetic biology. As we have gone through an exponential technological and scientific advancement, we have progressively abstracted ourselves from the natural and physical conditions imposed on us by nature, recreating narrowly controlled and regulated artificial environments around us. Today, the majority of the world's population lives in urban areas. The United Nations has projected that the trend is expected to rise further and reach almost 70% within the next thirty years (United Nations 2018). At the same time, people spend more than 80% of their lives indoors (Armijos Moya et al. 2019). As expressed by John M. Culkin: "We shape our tools and then they shape us" (Culkin 1967). This insight is proving to become ever more relevant, whether we discuss digital media or the built environment. The quality and the nature of the tools we inhabit and that make up our everyday environment impact not only our lifestyle and its sustainability, but also our physiological and psychological well-being.

One can argue that the capacity of a man-made environment to create variety in space and variability in time—which are main characteristics of the natural world (Baker 2000)—is not only an ecological and sustainability obligation, but also a fundamental human need in terms of opportunities to make adaptive responses. Can we provide a built environment with more complexity and richness, without falling back into a purposeless and simulated imitation of nature?

Change in Man-Made Things

It is a widely accepted principle in the field of evolutionary biology that the generalization of body features is a selective advantage rather than a flaw. In this context, optimal performance perfected through specialization is sacrificed, favouring the flexibility that derives from generalization. As an example, the rise of *Homo sapiens* as a species of generalists has been suggested to derive from the versatility and capacity to adapt (if only adequately) to all kinds of environments on the globe (Williams 2013). In the same way, the human manual skills have been read as the consequence of the developed bigger brains and coordination skills, forming them as the tools of a generalist (Cohen 2010).

If we translate these evolutionary principles into a man-made context, it is easy to notice how much we tend to perfect and control the design of the things we create. The modern society, so strongly relying on behaviours of consumerism and accumulation, drives a technological optimization (in this case, a synonym of overspecialization) that is strongly monofunctional, negating from the start the possibility for any change or spontaneous process to occur. Take a modern car, which can swiftly outperform a horse in terms of speed and power but requires a great amount of infrastructure to be able to function—from smooth roads to refined sources of fuel. The horse, on the other hand, is certainly less performing but swifter in adapting to different types of terrain (sand, mud, rocks, steep slopes, even to water by swimming) and easily finds sources of nourishment.

This kind of obsolescence embedded in our designs derives from intrinsic limitations in the design process itself, which mostly derive from a top-down mentality of the creator(s) who tend to shape an idealized design, overly controlling the development of the product in terms of:

- Shape and use. The product is either (i) meant to fit all (needs or users), failing to adapt to specific conditions or (ii) over-customized, becoming immediately obsolete as the circumstances change;
- Development in time. The authors seek to foresee change of use or purpose in time, limiting in fact the range of adaptation.

From an engineering and design point of view, adaptivity in artefacts can be achieved on two different scales: on a micro- and on a macro-scale.

In the first case, *microscale flexibility* acts through the engineering of materials and of their behaviour on a molecular and nanoscale. This field of expertise mainly pertains to chemistry, material engineering, biology and genetics. Once embedded within the matter, this intrinsic flexibility can be transferred to *macro-scale* products by integrating these specific substances that can augment their capacity to adapt.

In the second case, *macro-scale flexibility* happens on the design scale assembling and shaping different parts to produce our tools. These can range from the design of a microscopic robot to that of a building or a system of buildings. A great number of fields of expertise are touched by this topic, from most fields related to engineering such as product design, architecture, mechanics to more art-oriented contexts such as fashion, advertising and craftmanship. Macro-scale flexibility can be achieved through the design of the product and its parts in mainly three ways:

- through open and open-ended design;
- through disassembly;
- through motion.

This book addresses adaptivity on macro-scale and more specifically adaptivity of man-made things through motion. This is a kind of flexibility that responds to immediate needs and pressures: reactions happen within hours to seconds and are a fundamental functional part of the tool.

Open-ended design and disassembly on the other hand involve a hibernated kind of flexibility that can be activated when needed, on an occasional basis, to prolong the use of the system or of its parts over a longer timeframe. Although these concepts have radically different requirements than flexibility through motion, they are potentially complementary, which is why it seems worth to discuss them briefly.

Adaptivity Through Open and Open-Ended Design

Open design (OD) products are developed within a vision of free distribution and use of the design outcome, hence also allowing changes and evolutions of the original project. It is a logic that can be associated with the principles behind Do-It-Yourself (DIY) design. In comparison to the traditional design-production-use system, the OD shifts the paradigm towards the users becoming directly involved in the design and the production, also requiring specific skills and knowledge to allow the capacity and motivation to intervene in the design.

The Open-ended Design (OeD) philosophy on the other hand can be closely associated with the Japanese *wabi-sabi* aesthetic orientation, which sees beauty in the imperfect and transient nature of things. An OeD is usually characterized by the intentional and meaningful imperfection, incompleteness of its design features. In other words, it has many similarities to an unfinished project, which enables other actors to fulfil the missing requirements by adapting the OeD to a specific context in a different timeframe. End-users are supported by the design itself in their exploration and investigation of the object's functionality and operation relying on their own creativity and capacity to find new applications. Apart from having obvious implications in terms of long-term sustainability, OeD has the potential to accommodate change on various scales opening up possibilities for emergent behaviour, as well as to foster the development of additional values such as personalization and emotional bonds between the user(s) and the tool.

For the designer, creating an OeD by embedding imperfection or incompleteness does however not mean that the project is in any way lacking or poorly performed (Gaver et al. 2003). On the opposite, ambiguity has to balance randomness and accuracy very well, hinting at possible new interpretations without imposing any choices (Ostuzzi et al. 2017). The designer is strongly involved in choosing which aspects should be allowed to change and to which extent. The potential to change is constantly present, reflecting a spontaneous ability to support potential unpredictable change, rather than forcing change when the context of use does not require it.

Typical examples of OeD are platforms that provide an infrastructure allowing the users to implement it or create their own ideas and projects. This is the case of Wikipedia, which starts as an empty box, grows and dynamically perfects itself in time through its use.

An OeD that uses its imperfection as a statement of its nature is the "Do add" chair, where one leg is shorter than the other three, making it impossible to sit on it. Its function completely relies on the creativity of the user to take action and rebalance it.

In a similar way, the "Incremental Houses" projects of Elemental/Alejandro Aravena introduce a half-finished housing typology by building only one half of the intended volume and leaving the other half completely free for the users to fill. This allows not only flexibility and change but most and foremost a personalization and a re-appropriation of the house by the owners.

The LEGO brick is a highly versatile toy relying on a modular system enabling the users to build complex and unique shapes. The brick in itself is an incomplete design in the measure that one unit alone does not serve almost any function, but relying on the additive design of many units though, the function can change to please the user.

Adaptivity Through Disassembly

Efficient design for disassembly (DFD) is an effective method *reversing the assembly process* to reduce waste, simplify recycling, enhance maintainability, repair and substitution, as well as provide ready components for future re-assembly of other products (Barsan et al. 2007). Reversing the assembly is in most cases a difficult task which requires products to be designed for it. In a world of growing population and consumption with finite resources, DFD is however becoming of critical importance to prolong the usage of all our products, as they quickly become obsolete and are exchanged for newer and more attractive devices.

In this case, adaptivity is not found in the capacity of the product itself to change, but as a process during its overall life cycle. Ideally, the more the iterations of assembly–disassembly–reassembly, the longer the life cycle and the greater the adaptive capacity.

Adaptivity Through Motion

Most man-made artefacts are designed to move. In fact, kinetics is one of the most common strategies to embed adaptability in our tools: whether we want to redirect the light beam of our lamp, open a book or fold our eyeglasses to transport them better.

In these tools, their specific type of motion is often defining their function and usability: when motion is not possible anymore, these objects often lose their purpose and become waste. Unfortunately, moving parts in tools are also the most subject to wear and are likely to break. To contain those issues, the design of moving systems tends to finely balance the complexity of the kinetics in terms of mechanical design and the variation of tasks it addresses. It is not too surprising that the most basic kinetic tools we use have a high probability of also being the most long-lived ones—not only because they were easier to invent, but also because it is difficult to come up with a more efficient design for them. Less is more.

Change as a Transformation of Energy

Life and matter are all made out of energy. In physical terms, change can be described as the outcome that we can see and measure of the constant energetic exchange that happens between the synergetic systems interacting around, within and with us. The concept of synergy explains natural systems as working through economic principles of material and energy efficiency, acting on and interacting with given over-ordered and subordinated synergetic systems (ecosystems, organisms, etc.), determining each other's development by exchanging energy under different forms (Fuller and Applewhite 1982). An ecosystem can be described as a balanced system of constant energetic exchange between its parts.

Photosynthesis is the greatest chemical operation on the earth, transforming solar radiation into about 300 billion tons of sugar per year. And from this process, derives all the organic matter—from vegetal to animal organisms, including natural gases and petrol—that we have been consuming since we started evolving into our present form (Benyus 2002). We are only now slowly realizing how energy-inefficient we are, burning the inherited rests of trapped ancient sunlight (petrol) while ignoring smarter ways to transform energy. Whether we like it or not, we are bound to the ecological laws, which means that we have not secured the future of our species at all, although we might be now at the top of the food chain. In fact, the real survivors on the planet are those who find the way to live for generations over millions of years without consuming their ecological capital (Benyus 2002).

In the last centuries, we have combined unlimited growth in number with a very limited improvement in energetic sustainability—the complete opposite of what we should be doing. In nature, "more and more" works but lasts not, while "better and better" always works (Nachtigall 2000).

The Myth of Zero Energy

In physical and biological terms, the concept of any element with exact "zero energy" is rigorously meaningless. Living things must actively work to prevent death. When they cease to exchange energy, they cease to exist as autonomous beings and merge into the surroundings (Dawkins 1996). Energy and energy conservation, or energy optimization, are therefore key factors in all living organisms, that must adapt and convey the energy exchanged with their environment in terms of climatic, biological, mechanical and social factors in order to achieve balanced interdependent bonds for a sustainable endurance in time.

At the same time, concepts such as "zero energy (ZE)", "net-zero energy (nZE)" or "positive energy (PE)" are becoming more and more in use in many fields of engineering, and especially in the construction and building industry. These terms commonly refer to the "operational energy" (OE), which is the energy required during the entire service life of the building to ensure its functioning in terms of lighting, heating, cooling, ventilation, operating of building appliances, etc. Although common and unambiguous definitions of ZE/nZE/PE are still lacking, all these terms refer to different degrees in balance between the on/off-site renewable energy generation and the OE of the building. On the other hand, the embodied energy (EE), which encompasses the energy required in the life cycle stages other than the operational stage (Azari and Abbasabadi 2018) is largely disregarded—including the energy necessary to extract all materials, produce and assemble the components, transport, build, maintain the object and finally disassemble, reuse or dismiss it (Battisti, Persiani and Crespi 2019).

Although these low-energy qualifications of new buildings do have an important role in promoting the necessity to build with the operational energy consumption of the building in mind, it also risks diffusing some misunderstandings. It is therefore useful in this context to remind that:

1. For the end-goal of promoting sustainability in our built artefacts, the full life cycle behind the object must be taken into account, and therefore the operational energy and the embedded energy alike. Neglecting to design through a life cycle perspective can result in a problem shifting, where to solve issues concerning OE, more resources are spent upfront embedding more technologies and systems in the building at an expense of more EE.
2. Energy efficiency is not comparable to energy balance. Energy and resource efficiency refer to the concept that to perform a given task, the system is optimized so as to consume as little energy and resources as possible.

Energy balance, on the other hand, means that the energy produced (generally refers to energy from renewable sources) balances out the energy consumed. In other words, energy balance means that even an energy-inefficient building can potentially become "zero energy", provided that it can produce more energy than it consumes.

3. It is not possible for us to live, let alone do or create anything, consuming strictly "zero" energy (in the sense of no energy at all).

Technological Overshoot

One field in which high-tech solutions and systems are invested to achieve energy efficiency, often at a high cost of the EE is advanced building skins.

The building envelope, which comprises facades and roof systems, is the building system that has the largest surface in contact with the external agents. In the same way that the skin of a biological organism regulates energy and matter exchange between its internal and external environment, the building envelope has—with its multiple aesthetic and functional tasks—been recognized as a strategic surface for energy regulation, saving and collection (European Commission 2010). Building envelopes have therefore increasingly been the object of much technological research, with the aim to convert them from passive to active protective sheath regulators of energy balance (Luible et al. 2018).

In this new approach to improve the buildings' energy efficiency, a conventional approach to passive architecture has in many cases been coupled and enhanced with available technical advancements to achieve a higher grade of reactivity and control. The background concept is that to work effectively as an osmotic surface regulator and uphold the wished-for conditions inside the building, the building skin is designed so as to adopt different physical features (e.g. adaptable thermal transmittance) at different times of the day, in reaction to the conditions and the requirements. As a result, advanced building skins are designed to integrate technology at different levels to achieve environmental intelligence and autonomously regulate without the direct intervention of a human operator (Harris and Wiggington 2002).

The parallel rise of cheap and widespread technology made the use of technologically controlled kinetic devices not only possible in the field of architecture, but even quite widespread. However, the increased automation in building skins, while potentially being a clear advantage, is also associated with the main disadvantages of the technology it embeds:

- Intelligent building technologies that employ complex and expensive systems risk increasing the dependency on technology and energy rather than creating energy-saving solutions in the long run. The effort to produce and operate the technology is in this case disproportional to the performance it achieves.
- While technological ephemeralization increasingly enables us to do more with less (Fuller 1973), it also speeds up the rate at which our products are considered obsolete, becoming waste. The likelihood of the building technology

becoming obsolete is directly proportional to the amount and the complexity of high-tech components embedded in the design.
- Complex and high-tech solutions risk increasing defects in the operations and raising costs (manufacturing, installation, operation and maintenance) due to the lower interchangeability and compatibility of single components, as well as the potentially higher energy consumption derived from individualized usage behaviour (Gosztonyi et al. 2013).

From our global greenhouse gas (GHG) emissions, to the effects our activities have on the environment (Ritchie and Roser 2020), it would seem that—although we have reached unforeseen achievements—we have set a number of priorities wrong. The building industry alone is responsible for over one-third of final energy consumption and greenhouse gas (GHG) emissions globally (European Commission 2019), while textile production is responsible for more emissions than international flights and maritime shipping combined (Ellen MacArthur Foundation 2017; Nature 2018). Moreover, although being in an economy where change is the only constant, our products are designed and produced to outlive our present needs, impacting many generations to come (Sassi 2006). Therefore, unless we want our resources to end up in landfills, designing for "longevity" needs to assume a new meaning in terms of adaptability on short and on long term within a resource-efficient circular-economy mindset (Battisti, Persiani and Crespi 2019).

The Resources of Man-Made Design

So, where do we start in changing the way we relate to our environment? Although we might realize that we have done a lot of things wrong, we have also developed along the way a series of capacities and abundant knowledge which are extremely useful. If the problem is not the method in itself but rather the aim, we can treasure what we have learned along the way while we figure out how to rethink our goals and priorities.

What are the resources we have on our side, as designers, engineers, artists and creators?

Mother Nature as a Template

Man-made systems have often been developed in distinction or even in opposition to the natural environment. On the other hand, biological organisms have also been an enormous inspiration pool for developing new designs. In fact, nature exhibits an extremely refined use and combination of the available resources, evolving an enormous variety of systems and designs in an almost perfect energetic balance with their environment. So why not use the already existing and over-tested blueprints?

Using nature as a template relies on the human capacity for observing, investigating, mimicking and ultimately reinventing existing solutions in the environment.

Designing by Replicating

In designing by replicating, or copying, solutions and designs from a natural matrix are imitated or reinterpreted into an artificial context, to solve specific technical challenges. This design logic named after these principles is the biomimetic approach.

This is, for instance, the case of many advanced materials with particular proprieties that are the outcome of the observation of natural structures on a microscopic level which are successively reproduced through advanced nanotechnology.

An example is the nylon-based fabrics integrated with drag-reducing water-repellent features that mimic on the material's nanoscale the surface morphology of a sharks' skin, achieving an improvement of glide through water with a 38% reduction in resistance (Persiani and Battisti 2018).

Designing by Applying

In designing by applying, man uses nature as a co-worker, developing new techniques to craft materials in a way that is more similar to gardening and farming than to manufacturing (Collet 2013). Natural substances and organisms are used to replace artificial ones, with the benefits and the drawbacks that come with using living matter as a substance or a code: mainly irreversible processes such as biodegradability and decaying; multiplication, growth and healing. In this case, the main drawbacks can also be seen as interesting opportunities for rethinking innovative solutions.

An example is how MIT researchers used live silkworms to grow a small pavilion out of natural silk filament, using the caterpillars as source of material and workforce, rearranging the filaments according to a hexagonal shell framework (Oxman et al. 2013).

Designing by Combining

In designing by combining, artificial and natural materials or solutions are combined in order to strengthen each other's specific assets or features.

This is, for instance, the case of HygroScope: Meteorosensitive Morphology, one of the first humidity-sensitive and responsive surface installation. The surface is built out of multitude triangular composite flaps. The material is realized through a layering of maple veneer and synthetic composite triangles with programmed fibre direction, length, thickness and geometry in order to maximize the swelling and lengthening of the material in reaction to a rising humidity rate (Achim Menges Architect et al. 2012).

Designing by Manipulating

In designing by manipulating, biological matter is being reprogrammed (as in genetically modified organisms, or GMOs) or even rewritten from scratch using basic genetic building blocks (synthetic biology, or Synbio). The aim in this case is to reconfigure the living organisms to create man-made variants with pre-programmed features in order to perform specific tasks with a predicted outcome (Persiani and Battisti 2018).

Scientists have implanted fully functional bacterial luciferase within a tobacco plant to create prototypal specimens of GMO light-emitting plants with possible future applications in the lighting industry (Krichevsky et al. 2010).

The Sapiens Potential

Other resources available to man-made design, that integrate the nature-inspired copy/change/paste approaches, are found within the capacities that our species has evolved—from its social structure to the development of our most important tool: the brain.

The "Ratchet Effect"

Something that has been recognized as special in the way our species is creative—in comparison to other species—is that our creative ideas build on each other's cumulatively. This characteristic is called the "ratchet effect": a capacity of repeating, modifying and transmitting improved versions of a specific skill or habit over generations of individuals and large groups (Tennie, Call and Tomasello 2009). One generation does things in a certain way, which is repeated by the next generations in the same way until a new generation "puts a spin" on the inventions of others (Gabora and DiPaola 2012). Once an improvement is introduced, the optimized solution is taken as the new baseline and further transmitted to the next generations.

This process allowing to accumulate knowledge over time relies on a balance between (Tennie, Call and Tomasello 2009):

- inventiveness—which relies considerably on two other aspects that our species is very good at: observation and abstract thinking;
- ability to communicate our findings to a broader group;
- capacity to transmit knowledge across generations—regulated through processes of teaching, social imitation and normativity (the "right" way of doing things) to keep the novelties in place until other novelties come along.

This over-tested method of cultural development might be one of the main pillars standing behind the rise of improbable designs that survive over multiple generations and can be actively used as a design method, reinterpreting existing solutions.

The Associative Capacity of Abstract Thinking

Although the ratchet effect—on which human knowledge and technological development are based—fundamentally relies on an evolutionary methodological matrix, there is more to man-made innovation than a cumulative selection process. Unlike natural evolution that is restricted by the physical laws and space, human-driven innovation is driven by design in conceptual space, enabling us to jump over otherwise necessary evolutionary stages to achieve a specific solution.

Through cumulative selection, evolution has not necessarily designed the most efficient solution to one context, but rather an accumulated outcome of a series of solutions to specific stressing factors in time (Persiani 2018). This implies that evolution relies on a historical baggage that strongly influences the matrix of the design. The shape development process (or morphogenesis) of a zygote's (fertilized egg) body plan will for instance transit through successive stages reliving its evolutionary process from a more primitive organism into the final organism (Manuel 2009).

Technology, on the other hand, answers a specific situation in time with a design specifically tailored to that one problem. As such, it is an assembly of different parts that have first been developed as an abstract model within a specific framework of assumptions where even the physical environment can be reshaped.

A typical example of the unique morphological and functional design deriving from this method is the wheel, of which no counterpart has been found in nature up to today. Among the suggested explanations for this "technological leap" achieved by man over Mother Nature is that the wheel as an invention is too complex and requires too many specific boundary conditions to be useful: it must work perfectly to work at all. A wheel requires a complex wheel-and-axle system, flat roads and the entire cart to be built to be of any value at all (Anthony 2007). Hence, a rudimentary proto wheel would provide no evolutionary advantage whatsoever to any organism. Moreover, the necessity of having the wheel rotating freely around the axle poses a number of issues from a physiological point of view for the exchange of nutrients and waste of the cells for processes as growth and healing (Dawkins and Wong 2005).

The capacity for engaging in an abstract stream of thought, imagining the unreal and building an alternative model or vision of the external world by configuring solutions together has proven to be our greatest evolutionary asset so far. On the other hand, it has also brought us to abstract ourselves from the natural and physical conditions imposed on us by nature. Since we are as all other organisms bound by the ecological laws, we need to integrate our innate capacities for abstract matchmaking with the smart ways of Mother Nature, working along rather than competing against her.

The Book

This book aims to gather and categorize different aspects describing adaptivity through motion in man-made systems in an inspirational catalogue of kinetic features aimed to assist engineers, architects, designers and artists to integrate auto-reactivity in their work. The book provides not only a classification of important parameters and potential alternative solutions, but also schemes to identify those aspects with a high probability of working together efficiently.

In ▶ Chap. 1, the concept of autoreaction is presented and discussed as a new categorization of man-made artefacts that, partially, already exists. Autoreactivity is compared and differentiated from neighbouring concepts, such as kinetics, intelligence and interactivity. The fundamental hypothesis behind the creation of this new category is that autoreactive devices are, for a big part, an optimized version of existing kinetic devices. This optimization lies in finding the optimal fit between the design parameters defining the system: from the geometry to the choice of latent energy. The following chapters therefore discuss, one at a time, the main aspects in the design of an autoreactive device, building on the existing knowledge in each specific field but also introducing and borrowing theories, methods and inspiration from closely related disciplines.

The shape formation, or morphogenesis, of the parts involved in a movement has a huge impact on the potential of the kinetic outcome, and therefore also on autoreaction. ▶ Chapter 2 explores different potentials, linked to different types of geometries embedded in the artefacts. Inspiration is also taken from geometrical and biomechanical studies on biological organisms.

The mechanical aspects of autoreaction are discussed in ▶ Chap. 3. Mechanical and biomechanical principles are brought together to present an abstracted and simplified review of these principles to use for the optimization of the design of autoreactive systems.

The energy source is a most fundamental aspect in the autoreactive context. ▶ Chapter 4 analyses the potential latent energy sources at the base of the concept.

The substance making up man-made systems has a great potential to vehicle energy transformation and movement. In ▶ Chap. 5, the next generation of materials that can be used to transform environmental change into physical motion is discussed.

Existing design solutions sometimes already make a smart use of the available energetic resources to convey autoreactive kinetics, while others have the potential to be converted to autoreactive systems. ▶ Chapter 6 explores through a few case-studies the autoreactive potential of existing solutions. In this context, powered and unpowered kinetic devices are taken as example and discussed.

References

Achim Menges Architect, ICD University of Stuttgart, Transsolar Climate Engineering Stuttgart (2012) HygroScope—meteorosensitive morphology. Centre Pompidou, Paris

Anthony DW (2007) The horse, the wheel, and language: how bronze-age riders from the Eurasian steppes shaped the modern world. Princeton University Press, Princeton, NJ. ISBN: 978-0-691-5887-0

Armijos Moya T, van den Dobbelsteen A, Ottelé M, Bluysse PM (2019) A review of green systems within the indoor environment. Indoor Built Environ 28(3):298–309

Azari R, Abbasabadi N (2018) Embodied energy of buildings: a review of data, methods, challenges, and research trends. Energy Build 168:225–235

Arthur W (1997) The origin of animal body plans: a study in evolutionary developmental biology. Cambridge University Press, Cambridge, UK

Baker N (2000) We are all outdoor animals. In: Architecture, city, environment, proceedings of PLEA 2000, Cambridge, UK, pp 553–555. ISBN 1-902916-16-6

Baker N, Standeven M (1996) Thermal comfort for free-running buildings. Energy Build 23:175–182

Barsan L, Barsan A, Bolos C (2007) Design for disassembly as an essential step in fulfilling the eco-design goals: reuse, repair, re-manufacture, recycle. JIDEG J Ind Des Eng Graph Issue 3(3):22–24

Battisti A, Persiani SGL, Crespi M (2019) Review and mapping of parameters for the early stage design of adaptive building technologies through life cycle assessment tools. Energies 12(9). ► https://doi.org/10.3390/en12091729

Benyus J (2002) Biomimicry: innovation inspired by nature. Harper Perennial, New York

Cohen YA (2010) Human adaptation: the biosocial background, 1st edn. (n.d.), AldineTransaction

Collet C (2013) En vie/Alive. Alive: New design frontiers. Exhibition. Espace Fondation EDF, Paris [Internet]. Available from: ► http://thisisalive.com/about/. Accessed 14 Apr 2018

Culkin J (1967) A schoolman's guide to Marshall McLuhan. Saturday Rev 51–53:70–72

Dawkins R (1996) The blind watchmaker, why the evidence of evolution reveals a universe without design. W.W. Norton & Company, Inc. New York. ISBN 0-393-31570-3

Dawkins R, Wong Y (2005) The Ancestor's Tale: a pilgrimage to the dawn of evolution. Houghton Mifflin Harcourt, ISBN 9780618619160

Ellen MacArthur Foundation (2017) A new textiles economy: redesigning fashion's future. Retrieved on 21 Feb 2020 from: ► http://www.ellenmacarthurfoundation.org/publications

European Commission (2019) Energy efficiency, buildings. EU Policies, Information and Services. Available online: ► https://ec.europa.eu/energy/en/topics/energy-efficiency/energy-performance-of-buildings. Accessed on 18 Mar 2019

European Commission (2010) Directive 2010/31/EU of the European Parliament and of the Council of 19 May 2010 on the Energy Performance of Buildings. Official Journal of the European Union, Brussels, Belgium. Available online: ► http://eur-lex.europa.eu/LexUriServ/LexUriServ.do?uri=OJ:L:2010:153:0013:0035:en:PDF. Accessed on 2 Aug 2017

Fuller RB, Applewhite EJ (1982) Synergetics, explorations in the geometry of thinking (n.d.). Macmillian, New York

Fuller RB (1973) Nine chains to the Moon. Cape, London

Gabora L, DiPaola S (2012) How did humans become so creative? A computational approach. In: Maher M, Hammond K, Pease A, Pérez y Pérez R, Ventura D, Wiggins G (Eds) Proceedings of the third international conference on computational creativity. Association for the Advancement of Artificial Intelligence, Palo Alto, CA, pp 203–210. ISBN 978-1-905254668

Gaver WW, Beaver J, Benford S (2003). Ambiguity as a resource for design. In: Proceedings of the 2003 conference on human factors in computing systems, CHI 2003, Ft. Lauderdale, Florida, USA, 5–10 Apr 2003. ► https://doi.org/10.1145/642611.642653

Gosztonyi S, Gruber P, Judex F, Brychta M, Richter S (2013) BioSkin—Forschungspotenziale für bionisch inspirierte energieeffiziente Fassadentechnologien, [Research potential for bionically inspired energy-efficient facade technologies]. Berichte aus Energieund Umweltforschung 46/2013, Nachhaltig Wirtschaften, Bundesministerium für Verkehr, Innovation und Technologie

Harris J, Wiggington M (2002). Intelligent skins, Elsevier LTD, Oxford. ISBN 978-0750648479

Krichevsky A, Meyers B, Vainstein A, Maliga P, Citovsky V (2010) Autoluminescent plants. PLoS One 5(11):e15461. ► https://doi.org/10.1371/journal.pone.0015461

Luible A, Aelenei L, Perino M, Wellershoff F, Pottgiesser U (eds) (2018) FAÇADE 2018—Adaptive!; special issue COST action TU1403 adaptive facades network. J. Facade Des Eng 6(3). ► https://doi.org/10.7480/jfde.2018.3

Manuel M (2009) Evolution/Évolution: early evolution of symmetry and polarity in metazoan body plans. Comptes Rendus—Biologies, n.332 (La théorie de Darwin revisitée par la biologie d'aujourd'hui/Darwin's theory revisited by today's biology), pp 184–209

Nachtigall W (2000) Das Große Buch der Bionik, Neue Technologien nach dem Vorbild der Natur, Deutsche Verlags-Anstalt, Stuttgart/München

Nature (2018). The price of fast fashion. Nat Clim Change Editorial 8(1). ► https://doi.org/10.1038/s41558-017-0058-9

Ostuzzi F, De Couvreur L, Detand J, Saldien J (2017) From design for one to open-ended design. experiments on understanding how to open-up contextual design solutions. Des J 20(sup1):3873–3883. ▶ https://doi.org/10.1080/14606925.2017.1352890

Oxman N, Keating S, Kayser M, Duro-Royo J, Gonzalez Uribe C D, Laucks J. MIT Medial Lab. (2013). Swarm Printing, Silkworm Silk Deposition [Internet]. 2013. Available from: ▶ http://matter.media.mit.edu/tools/details/Swarm-Printing [Accessed: Apr 9, 2018]

Persiani SGL, Battisti A (2018). Frontiers of adaptive design, synthetic biology and growing skins for ephemeral hybrid structures, In: Membrane technology—energy efficiency and sustainability, InTechOpen. ▶ https://doi.org/10.5772/intechopen.80867

Persiani SGL (2018) Biomimetics of motion, nature-inspired parameters and schemes for kinetic design. Springer International Publishing, Cham, Switzerland. ▶ https://doi.org/10.1007/978-3-319-93079-4. ISBN 978-3-319-93078-7

Ritchie H, Roser M (2020) CO_2 and greenhouse gas emissions. Published online, OurWorldInData.org, University of Oxford, Global Change Data Lab, Retrieved from: ▶ https://ourworldindata.org/co2-and-other-greenhouse-gas-emissions

Sassi P (2006) Strategies for sustainable architecture. Taylor & Francis Inc., New York, NY, USA. ISBN 9780203480106

Schein EH (2002) Models and tools for stability and change in human systems. Reflections SoL J. ▶ https://doi.org/10.1162/152417302762251327

Tennie C, Call J, Tomasello M (2009). Ratcheting up the ratchet: on the evolution of cumulative culture. Phil Trans R Soc B 364:2405–2415. ▶ https://doi.org/10.1098/rstb.2009.0052

United Nations (2018) 68% of the world population projected to live in urban areas by 2050, Says UN. In News; United Nations Department of Economic and Social Affairs, New York, NY, USA, 16 May 2018. Available online: ▶ https://www.un.org/development/desa/en/news/population/2018-revision-of-world-urbanizationprospects.htm. Accessed on 13 Feb 2020

Williams C (2013) Origins of form, the shape of natural and manmade things, 1st edn. Taylor Trade Publishing, Lanham

Contents

Autoreaction

Contents

S. G. L. Persiani, *Design of Autoreaction*,
Design Science and Innovation, https://doi.org/10.1007/978-981-15-6178-8_1

Technological advancement goes along with the evolution of human society. It is indicative of the knowledge and the manufacturing skills possessed, as it mainly serves specific purposes of support to the user's activities. A more efficient, precise tool can spare time, strength—and not less—energy involved in its operation. It is also indicative of a society's culture, the aims and the vision it pursues, as well as the economic power invested in developing specific types of tools.

Humans typically set the personal and social aims within timeframes relating to their lifetime and hence very short timeframes in comparison to ecologic and natural rhythms. It then becomes understandable that human societies result in putting enormous amounts of resources to develop technologies which in many cases pursue dangerously short-sighted advancements.

Humanity has a long history of pursuing promises and visions for a better future, in response to the present-time challenges, not least in terms of technology. We then realize looking back, how many wrong turns we took and how narrow-minded we were. We realize, 200 years later, that the industrial revolution that would have allowed human technology to surpass nature in about anything, brought many long-lasting negative drawbacks that we now need to tackle—among which a consistent rise in population combined with highly polluting and resource-consuming manufacturing and transportation technologies (Benyus 2002).

Something all these visions have in common is a lack of vision over a broader timescale. Even when a larger timescale is looked at, the immediate benefits are often positively overestimated in comparison to the long-term drawbacks. Design processes need to integrate visions over long timeframes, taking into account:
- whole system in which the solution will be used;
- the impacts on a global context;
- both technical and non-technical issues synergistically;
- multidisciplinary expertise;
- strive to solve the problem for a potentially infinite future.

1.1 Designing Change

Total adaptability is very hard to achieve: an object which can adapt to any kind of shape can have no original shape. Reversely, the more an object varies to solve all issues, the less it satisfies specific needs or aims (Fox and Kemp 2009). This issue is typical of many architectures, where the authors' conscious of the growing obsolescence of their works in time has attempted to forecast the future in two main ways, typically failing:
- designing specific solutions for supposed or imaginary problems, that possibly never occur;
- designing universal spaces that can potentially meet any requirement and ending up addressing none.

Designing the transition in time rather than single solutions, whether on large or on small scale, might be one possible approach to develop new solutions. While many engineers, architects, designers tend to plan solutions as if frozen in time, we have to realize that such a condition is mostly an abstract product of our minds than an accurate representation of reality. The real environment we live in is in constant change and sometimes subject to unforeseeable changes. As life within a building does not fully start the moment the structure is ready, it also does not crystallize within one form, neither in the present nor in the future.

Today, one of the main driving forces for developing new technologies and solutions are the changing human patterns of interaction with the built environment (Zuk and Clark 1970).

Kinetic systems introduce concepts of cause and effect (Jormakka 2002) with no single representative look in time as the geometric design patterns shift constantly (Moloney 2011).

In most contexts, the results of a spatial adaptation in time accommodate one or more of the following three aims (Fox and Kemp 2009):

- subjective experience-related adaptation, where the content or the message is dynamic and interactive transforming both the space and the user itself;
- functional adaptation, where the system is transformed by addition or subtraction of parts, following the users' temporary and immediate needs;
- performance-based adaptation, where the system reacts within cyclical patterns to answer changing operation needs.

1.1.1 Content Adaptation

Moving parts can be used to communicate and transmit change on completely different levels than the mere physical one.

Change in degree expresses motion through a quantitative differentiation of the elements involved. Variations within the single parts is potentially infinite but is limited to one dimension (Moloney 2011). An example is the "Wind Silos" (◘ Fig. 1.1), a screen composed of thousands of metal sheets which can tip around a fulcrum, changing orientation as the breezes blow through the screen. Although the possible variations of the sheets on the façade are potentially infinite, the human mind can conceptually grasp all these variations as they are limited to the morphospace occupied by the façade. Summarizing, change in degree is characterized by homogeneous units, juxtaposition in space, succession and quantitative representation (can be described by a number).

Change in kind expresses a qualitative differentiation of the elements involved. In this case, the interaction between parts can produce evolving and changing patterns with communicating different depth, complexity and unpredictability (Moloney 2011). An example would be a split flap display board (◘ Fig. 1.2) found at any railway station. In this case, the variation between the letters on the billboard is secondary to the message that can be communicated: the change is in this case limited in physical terms but potentially infinite on a psychological and

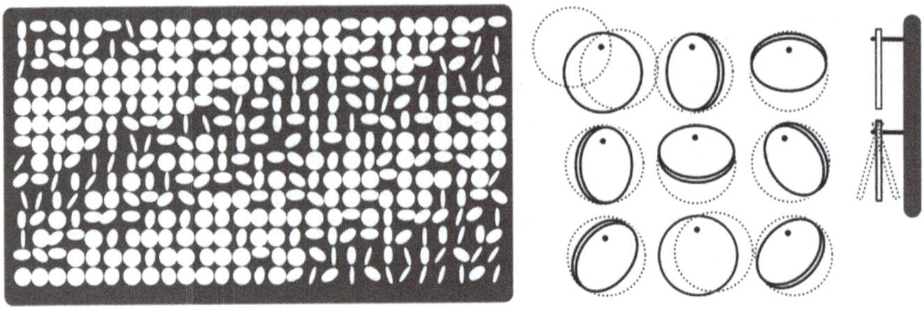

Fig. 1.1 Variations in degree, example from the "Wind Silos", International Trade Center by artist Kahn (2006). The stainless steel screens that cover the silos are composed of thousands of wind-activated, 15-cm-diameter stainless steel discs. Left, overall effect of the variations on the screens; right, scheme of the movement range of the single metallic discs

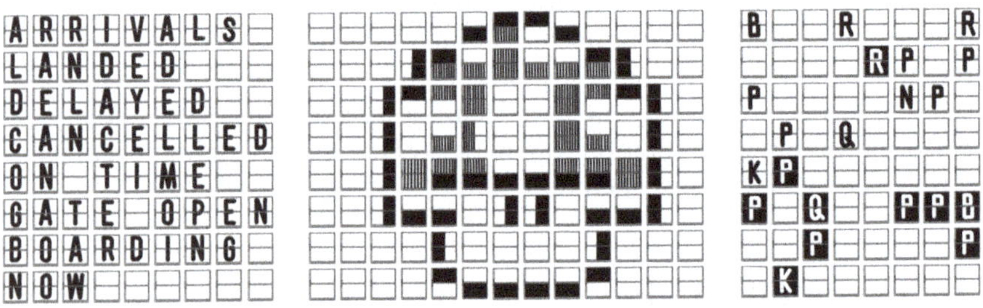

Fig. 1.2 Variations in kind, example from a split flap display board with different possible messages: (from the left) text message, image with possible cultural associations, representation of a process—in this case the 28th move within the chess game Kasparov versus Topalov

subjective level. Summarizing, change in degree is characterized by heterogeneous units (different letters), fusion in space (letters grouped into words), continuity (forms a message) and qualitative representation (can be described by a concept).

1.1.2 Incremental Adaptation

Incremental architecture is an open system which can integrate new external elements that did not exist at the time of the creation of its original shape, by adding, subtracting or substituting material. The system is not connected to a specific shape or form but is transformed in time depending on the context. The configuration must therefore not necessarily be defined precisely from the start, which makes it possible for these systems to respond to a wider range of issues than other kinds of systems (Zuk and Clark 1970). In that sense, the concept is almost closer to an evolutionary and biological concept of growth and decay, developing over a long timeframe of months, years and decades.

Adaptability is essential to the functional durability of a concept, especially in a contemporary context where customers' needs, and habits evolve at an increasing speed. Incremental adaptation can be found within kinetic man-made artefacts used to make static solutions more pragmatic. These moving components enable us to accommodate specific ranges of activities as they change geometry or orientation (Oosterhuis 2012). The moving parts add or integrate manually adjustable, modular and easily repairable or replaceable structures (as doors, windows, panels, etc.). These solutions introduce change, liveness and a broad range of other benefits that typically follow—aesthetics, function, flexibility, reliability, user friendliness, energy use, lifecycle costs—which have for a long time allowed to give better answers to functional problems than many present-day high-tech systems (May 2010).

1.1.3 Cyclical Adaptation

Cyclical change can be found in all systems with set outcomes, that, being controlled by a predetermined combination of factors are bound to continually repeat the same patterns, whether on very short or very long and more unpredictable timeframes. Cyclical change is therefore typical of mechanical and man-made solutions, where in most cases the same outcome is achieved over and over again without any development.

The flexibility and adaptability are in this case performance-based and driven by single events occurring throughout the cycle, from phase to phase until the cycle has been completed. Any self-optimization and adaptation of the system is therefore limited to single occasions in time, within the same overall cyclical pattern, with no potential change on longer timeframes.

This is the case of most kinetic systems that will be further explored in this chapter.

1.2 Kinetic Responsive Technologies

The increasingly widespread availability and accessibility to technological solutions has given rise to "technological hype" and rapid diffusion of kinetic devices into many contexts, from art and design to architecture—as to convert building envelopes into active, energy- and emission-saving osmotic surface-regulators to uphold stable indoor conditions (Harris and Wiggington 2002). The many technologies that are the outcome of this "hype" and of the accelerated search technological advancement can be described depending on their complexity and level of responsiveness.

In general terms, responsiveness in a kinetic system involves that the kinetic reaction is the outcome of a dialogue, or an exchange of information, originating from outside the system itself. This family of systems answer to immediate needs and changes, with physical changes that happen on very short timeframes, within seconds or minutes.

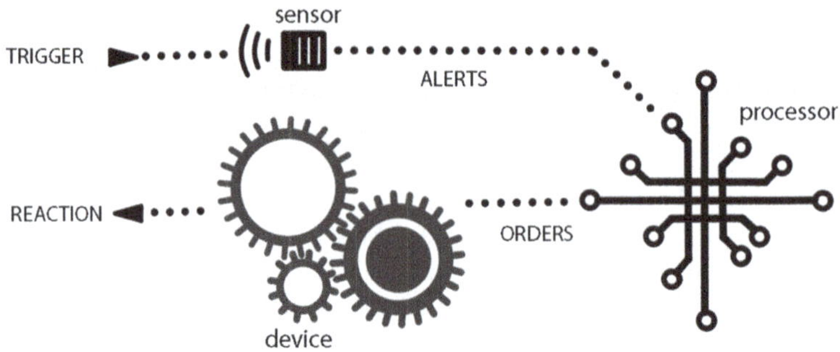

Fig. 1.3 Operating scheme of a responsive technology. An example of this type of system is the façade of the "Arab World Institute" in Paris is designed to react to the outdoor light intensity by filtering the incoming light through a system of motorized facade lenses able to adjust their aperture, controlled by a complex sensor-brain network

Responsive devices are centralized systems with a complete sensor-brain technology that measures a specific set of environmental conditions through sensors in communication with a central control system which in turn decides on adjustments of shape, colour or character through the command of actuators (Fig. 1.3). Specifically:

- The number of stimuli detected depends on the technology of the sensors integrated;
- The precision of the stimuli detection depends on the technology of the sensors integrated;
- The diversity between patterns of reaction is determined by the computing capacity of the brain element;
- The kinetic reaction is diffused over a large area;
- The kinetic reaction produces essentially a variation in terms of change in degree;
- Detection of change, processing and reaction are all powered actions.

Reactive or *adaptive* (Aelenei et al. 2015) devices are decentralized systems where the element responsible for both the sensing and the reaction is integrated within the same device (Fig. 1.4). The system is able to react mainly to one kind of stimulation and responds with a predetermined reaction (Moloney 2011). Specifically:

- The system reacts autonomously to mainly one type of stimulus. To achieve reactions to more stimuli, single reactive devices with different triggers need to be added;
- The precision of the stimuli detection depends on the design of the system and the technology embedded in it;
- The patterns of reaction are predetermined by the design of the system and recur always in the same patterns;
- The reaction has a local impact, affecting the area immediately adjacent to the device;

Fig. 1.4 Operating scheme of a reactive technology. An example of this type of system is the "Tessellate™" screens at the Simons Center for Geometry and Physics, Stony Brook (Zahner 2009). This sunscreen is made of a series of overlapping perforated metal screens that move on the same plane regulating its transparency

- The reaction produces a variation in terms of change in degree;
- Detection of change and reaction can be powered or also not, depending on the technology.

Adaptronic systems combine the characters of a centralized and a decentralized system, integrating a conventional centralized sensor-brain system and at least one functional material that allows localized embedded action and reaction to a specific stimulus. These technologies allow structures to be intelligently designed to respond locally to single stresses, avoiding the typical over-dimensioned designs calculated based on the worst-case scenario (Janocha 2007).

- The centralized system can react to as many stimuli as its technology is designed for, whereas the decentralized system reacts to one type of stimulus only;
- The precision of the stimuli detection depends on the materials or the technology embedded in the system;
- The patterns of reaction are predetermined by the design of the system and recur always in the same patterns;
- The reaction has a local impact, affecting the area immediately adjacent to the decentralized system;
- The reaction produces a variation in terms of change in degree;
- Detection of change, processing and reaction are mostly powered actions.

Interactive systems can have centralized or decentralized control. The peculiarity of these systems is in the reaction scheme which is a multiple-loop system allowing the device to adjust or "learn" from the precedent experiences giving a different output for a potentially identical stimulation (Fig. 1.5). Interaction involves a mutual and reciprocal action or influence between the technology and another system (Fox and Kemp 2009).

- The system can react to as many stimuli as it is designed for, combining multiple technologies;
- The precision of the stimuli detection depends on the materials or the technology embedded in the system;

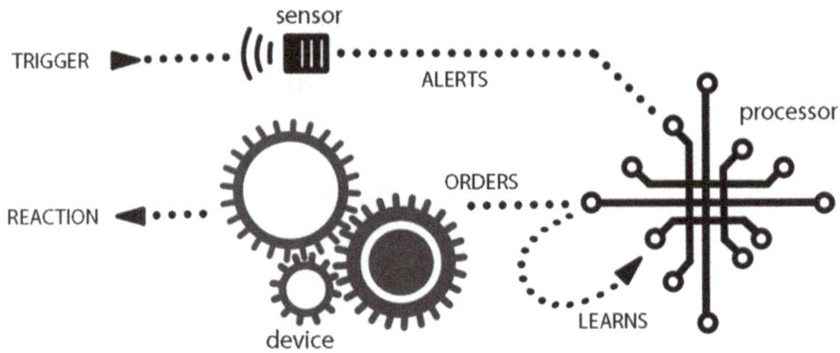

Fig. 1.5 Operating scheme of an interactive technology. An example of this type of system is the "Hylozoic Ground" sculpture series (Beesley 2010). Hylozoic Ground is an interactive sculpture, a forest of suspended dynamic geotextile plant skeletons, that respond to its surroundings by detecting movements and reacting with changeable movement patterns. In this case, a kind of conversation can be established between the visitor and the technology, which emulates a biological thought process (Harris and Wiggington 2002)

- The initial patterns of reaction might be predetermined, but the response can evolve upon use and become difficult to predict;
- The kinetic reaction can be local or diffused, depending on how the kinetic system is designed;
- The kinetic reaction produces a variation in terms of change in degree, but can potentially also produce a change in kind;
- Detection of change, processing and reaction are mostly powered.

From the solutions described above, it becomes quite clear that this technological development and "optimization" has greatly focused on the search for more dynamic and responsive solutions, and comparatively little on the progressive simplification and independence from advanced manufacturing and energy use. There are hence multiple gaps that can be identified in the kinetic technologies described above.

1. Complexity. More performing is not necessarily better. The range of complexity of the selected technology must be proportioned to its task;
2. Resource optimization. When intelligent, complex and expensive technologies begin to employ disproportioned amounts of resources in comparison to the achieved performance, then the true potential of technological ephemeralization has not been properly used (Persiani et al. 2016);
3. Purpose. Kinetics should not be researched for the only purpose of developing the upcoming technology, but motion itself must be given sense (Moloney 2011).

Technology is inevitably a double-edged sword. And technological progress generates both positive contributions to human welfare and negative environmental and social impacts. How should we then go about, to develop truly sustainable technologies, avoiding creating a number of new issues for every problem we try to solve?

Some authors imply that the path towards sustainable technological development lies in the maximizing the positive impacts while minimizing the negative ones. It is within this line of thought that autoreactive systems are introduced as an alternative within this family of kinetic technologies.

1.3 Autoreactivity

Autoreactive systems belong to the class of structures that aim to achieve a sort of technical homoeostasis, intended as an automatized regulatory reaction to a specific set of conditions. In that sense, autoreactive systems are a variation of reactive systems, in answer to the three gaps previously identified (complexity, resource optimization and purpose).

As highlighted in the definition above, the autoreactive nature of a system is identified by the co-existence of four main characteristics:

- autonomous reaction;
- kinetic output;
- leftover energy;
- functional purpose.

Other additional characters that can be found, sometimes recurrently, in autoreactive systems can be interpreted as extensions or refinements of the same concept. A few of these are:

- reaction to more than one input;
- embedded materials with autoreactive proprieties.

These features are, however, not considered essential to the overarching understanding of the notion of autoreactivity.

Definition
An autoreactive system is an autonomously acting system that uses available latent energy sources from its surrounding environment to achieve, for a set purpose, kinetic change.

1.3.1 Autonomous Reaction

Autonomous reaction focuses on a simplification on system level in answer to the question of technological complexity, involving the question of resource optimization as well.

The prefix *auto-* involves a self-trigged, predetermined answer to a stimulus without the intervention of any (internal or external) entity to direct the response. Hence, processes that perfectly function being automatically regulated and controlled are decentralized, cutting out the sensor-brain system that is otherwise used to achieve reaction (◘ Fig. 1.6).

In this context, autoreactive systems are considered as a technological parallel to the autonomic nervous system (ANS) that controls many of the everyday functional routine processes in our body: thermoregulation, automatic adjustments

Fig. 1.6 Operating scheme of an autoreactive technology. An example of this type of system is the "Hanging Mobiles" series of Alexander Calder (2015). The hanging mobiles are a series of metal plates, assembled through metal sticks, moving with two degrees of freedom and equilibrating each other. These twist and turn in thousands of variations according to the imperceptible air movements in the surrounding environment

of the eye muscles to light conditions, the cardiac beat etc. In fact, the human brain—as any man-made processor—is a high energy-consuming device that allows us to give complex and tailored reactions to different situations. As, however, not all processes in our body require such complex processing, nature has set priorities subdividing our nervous system into a high-energy complex processing system (the Somatic Nervous System, SNS) and an automatized low-energy system (the Autonomic Nervous System, ANS) (Harris and Wiggington 2002).

Apart from saving energy on processing effort, not having to process all reactions through a central control element frees up space for other functions. If on the other hand all normal routine reactions need to be processed through the central brain, additional capacity would be needed to process any additional critical high-priority decisions. Finally, cutting on processing effort also saves time from the communication between the peripheral areas and the central processor, allowing more immediate reactions.

Other inspirational examples can be found in biological systems, which are massively decentralized by nature. Many basic movements are achieved through automatized reactions in response to external change of conditions, without any apparent interference of a brain device. Examples vary from the hygroscopic kinetic proprieties of grass leaves, the rapid trapping movements of the carnivorous "Venus Flytrap" plant (Persiani 2018), to the capacity of each living cell to generate enough energy to sustain itself (Gibson et al. 2010).

There are many reasons for upholding automatized reactions in technology as well: technologic simplification, optimization, energy saving. While some uses require broad flexibility and the possibility for users to intervene in the adaptation sequences, many actions respond to complex but limited variations of inputs. One example is building skins that regulate the homoeostatic processes between indoor and outdoor environment. Prioritizing autoreaction would mean to save energy, materials, and to simplify the reaction patterns allowing faster reactions.

Obviously, not all technologies should be made to react autonomously. The advantages of using such a specific technology are highly context specific. Some technologies need broader flexibility and the possibility to actively control and modify the adaptation sequences. Choosing an autoreactive solution in such a context would give highly inefficient outcomes. This aspect is further discussed in ▶ Sect. 1.5. What however should be assessed is if there are any single parts or functions in a technology that can be integrated with or converted to autoreactive properties.

1.3.2 Kinetic Output

Motion can be described as a series of successive *changes in position* of a point within a chosen coordinate system (with origin, coordinate axes whether Cartesian, cylindrical or spherical, and unit scale quantifying distances) and defined through parameter of *time*.

An autoreactive system responds to a set trigger by producing movement. This is a fundamental aspect that distinguishes autoreactivity form many other forms of adaptive systems, which usually take into account all kinds of physical changes: optical, adhesion-changing, energy-exchanging, -generating or -storing capacities that can be integrated by using specific multifunctional materials (Ritter 2007).

Autoreactive kinetics generally follow the reaction patterns achieved through a chain reaction—which can happen either on macro-scale (as through mechanic transmission) or on microscale (as within a material's structure). The pattern of reaction is predetermined by the design of the system which typically produces one or a set of results for each stimulus. In that sense, the movement can be designed to fit a specific purpose, but variations of it are only possible in terms of degree.

An essential aspect in the design of the kinetic output is the *reversibility* of the movement. In order to have an autonomous device, the system must work in a closed-loop system where it is enabled to return to its starting position before the next reaction takes place. While in some systems, the initial shape is regained as soon as the trigger is removed (as when an elastic rubber band snaps back as soon as the force acting on it is removed), in other cases an antagonist mechanism must be designed to return the system to its original starting point (as when winding up a mechanical clock).

In terms of resource optimization, one of the main objectives during the design of the kinetic reaction concerns the maximization of the output for a given trigger. The physical magnitude(s) that are object of the optimization depend on the function of the system and are case-specific:
- velocity;
- acceleration;
- force;
- range of action;
- precision;
- fatigue;
- etc.

Depending on the aim pursued with the autoreaction, the parameters of main importance must be defined as well as the wished performance. In most cases though, the magnitude of the parameters will strongly depend on the intensity of the impulse providing the energy for the movement, although the upper and lower limits of operation will mostly be defined by the design of the system and the technology embedded in it.

1.3.3 Leftover Energy

Using leftover (or latent) energy to power the reaction is a concept that directly aims to provide a solution for the gap in terms of resource optimization. On one hand the aim is to make the most of what the present-time technological ephemeralization has to offer, and on the other to put the real energy need into a new perspective of "How much energy do we really need?" and "is there any opportunity offering an available energy source that can be used in its current form, which would otherwise be lost, rather than extracting or transforming it?"

Whatever activity we will be doing in the future, we will need energy under some form. Therefore, autoreactive systems are designed to respond to two challenges:

- Minimize the need of energy, achieved by optimizing the relationship energy input/output making good use of technologic ephemeralization and smart design;
- Use exclusively latent and unused forms of energy from the surrounding environment.

The principle at the base of the concept is that autoreactive systems divert unused and leftover energy from the surrounding environment, either through mechanical transmission or transforming other energy forms through specific material proprieties, turning the available form of energy into the wished kinetic energy—which is then used for a set purpose.

The different forms of energy that can be found as latent energy is a topic that is further developed in ▶ Chap. 4. Working exclusively with available waste energy deriving from other processes, all systems with a direct energetic input are excluded from this categorization. A bicycle is therefore not an autoreactive system, as the user must power the system in order for it to work.

The temporariness of these energy sources strongly characterizes autoreactivity. If on one hand, not having the energy sources always potentially available can give an impression of great unreliability, the opposite is true as well: if the system is well-fitted to the purpose, it will be activated only when it is useful. In contrast to most man-made interventions, autoreactive systems are not designed to resist to the ever-changing state of their surroundings but are designed exploit unbalanced states to adapt and respond to change.

1.3.4 Functional Purpose

As autoreactivity aims to be an optimization in terms of resources, the whole design of such a structure is strongly tailored on its function and on its part within the overall system. It is therefore logical that the scope of the reaction is an important and essential part in the development of the whole concept.

However, of all four main characteristics defining autoreactivity, the functional purpose is probably the most difficult one to establish—not least because the limits of the concept itself are subject to controversy and different possible interpretations.

Functionality is commonly associated with the practicality and usefulness of a specific solution, often referring to the way it works or operates, in serving a defined task (Oxford English Dictionary 2020). The idea of functionality is often presented in opposition to the concept of beauty and attractiveness, not least in the Vitruvian "*firmitas, utilitas, venustas*" (lat. stability, utility, beauty). As however the limits of the concept of usefulness are strongly dependent on the task, in many cases the presence of a functional purpose can be controversially argued for.

As reactive technologies are still under development and experimentation, many case studies that could be classified as autoreactive can in fact be considered as case limits, especially in the context of art installations. Does an artwork with autonomously moving parts have a purpose other than being beautiful? If art has a set task beyond simple beauty, we must conclude that also an artwork has a functional purpose and can hence be considered autoreactive. Other systems as windmills, are naturally much less controversial to categorize, having a very obvious function.

In this context, we can define additional limits to the usual concept of functionality. The purpose of an autoreactive system should not be limited to scopes that are an end in itself, as:
- achieving beauty;
- creating a movement.

1.4 Achieving Autoreactivity

The concept of autoreactivity is new in the measure that it has only recently (Persiani et al. 2015) been rationally separated from other notions of kinetic and adaptive systems. As mentioned previously, this does not mean that autoreactive systems have not existed in man-made systems for a long time already.

Autoreactivity in an artefact can be achieved (i) on a system level, (ii) on a material level or (iii) both.

(i) The autoreactive capacity is achieved on a system level. Non-dynamic parts form together a mechanism in which the interaction between the parts achieves the movement. If one or more parts are missing or do not perform correctly, the system as a whole might fall short of its autoreactive capacity;

(ii) The autoreactive capacity is embedded inside one or more materials composing the parts of the system. The parts of the system that do not have themselves any autoreactive potential, are generally aimed at supporting and emphasizing the autoreactive capacity of the material. In this case, the parts realized with autoreactive materials keep their reactive capacities regardless of the functioning or presence of the non-reactive parts;

(iii) The autoreactive capacity is a combination of the two previous cases. It is a mechanism combining non-dynamic mechanic parts and parts made of materials with embedded autoreactivity. If either the material or the systems' parts do not perform correctly, the system as a whole risk to fall short of its autoreactive capacity.

Autoreactivity on system level (i) is the typical case of a mechanical system that is made to move using a latent energy source, as in the case of a mechanical spring-scale that measures mass by reporting the distance that a spring deflects under a load. Autoreactivity on material level (ii) involves mechanics on microscopic scale, within a material. This is the case of human hair which, due to the hygroscopic proprieties of the hair's chemical structure, tend to curl and frizz in humid air conditions. Autoreactivity combining the mechanics and the material structure (iii) is the case of a hair-tension hygrometer that uses the hair's change in length under humid conditions magnified by a mechanism to measure relative humidity.

Regardless of the level on which autoreactivity is implemented, the key to a successful autoreaction lies in the optimization of all aspects involved in the design to achieve a maximum of movement with a minimum of effort.

By *reducing mechanical complexity*, focus is set on good design in the initial conception phase, using all available resources, rather than solving issues later in the design process using unnecessarily over-engineered solutions. Therefore, simple and smart solutions that make the most of the available technical and technological knowledge are prioritized. Simple designs in mechanical terms allow systems to become more reliable, reducing the number of interacting parts—and hence planning imperfections, maintenance and operating difficulties—but potentially also reducing the number of parameters to control.

Through a *multi-parameter integration*, the aim is to identify the best combination of geometry, motion type and transmission with the correct actuating system. Hence, select and assemble the best possible combination of form, materials, mechanics and energy to perform a specific function-oriented kinetic action in a resource-efficient logic. In this context, optimization is interpreted as the utmost "energy effectiveness" when material, form, function, structure and motion are strictly interdependent in a fit design combination.

1.5 Limits of Autoreactive Systems

In technology as in nature, optimization is not only about using as little energy as possible, but the overall energy balance must be convenient as well. Autoreactivity has, as all systems, its limits and flaws.

1. limits due to the context;
2. limits due to user acceptance;
3. limits linked to the source of energy;
4. limits due to the independence of the system;
5. physical limits of the system.

As any other technology, autoreaction must be used correctly and in the right context. Introducing autoreactivity in man-made artefacts is about integrating it in those layers which are potentially more efficient as autoreactive systems rather than converting whole systems into autoreactive ones. As previously mentioned, the evolutionary advantage of using an autoreactive solution is context specific. Some contexts require a direct energy input into the system to achieve a bigger flexibility and control, thus justifying the major energy requirements. A machine with high control and versatility needs as a computer has no chance to meet the needed standards, if designed autoreactively. Although such a system would work its entire life using no operational energy, it will always represent a waste of resources, time and effort unless it fulfils the set aims efficiently. In the overall energy balance of a kinetic device, the energy input must be considered as one among many factors to take into consideration.

Limits due to user acceptance can occur in autoreactive systems that are in close contact with users or interact/interfere with their behaviour patterns. Reactive and responsive building skins for instance have been found in different occasions to clash with users' needs and comfort, in terms of both indoor light regulation and comfort perception. Liveliness is not always appreciated by the users in buildings, giving them an uneasy feeling of not being able to control their environment (Gruber 2011). The movements and the light conditions recreated inside the Institute of the Arab World in Paris have been reported to be perceived as funny and disturbing for people spending a longer time in the building (Ritter 2007), while in residential buildings the users tend to override the automatically closing façade shutters prioritizing visibility and direct sunlight in the indoor living areas.

Limits to the system can also be brought back to the nature of the energy source, rather than to its temporary availability.

Risks linked to the temporary availability in time and quantity of the energy source are assessed as a relatively low; not because the source cannot undergo strong fluctuations, but because the purpose of the autoreaction is strongly connected to the specific conditions that also trigger the reaction. Hence, availability of the energy should coincide with necessity of reaction and vice versa.

Risks linked to an inaccurate matching of an energy source with the timeframes of reaction could potentially give rise to two possible situations:

- The autoreaction is not controllable by factors that have not been designed to trigger a chain reaction. Hence, if other factors than the selected ones recreate the environmental conditions that require a reaction, but the triggering element is missing, the device will not be able to react although there is the necessity for it.
- The reverse situation is also possible: an autoreaction is triggered although there is no actual need for it.
- The magnitude of the reaction is not calibrated to the triggering capacity. In other words, although the conditions for an autoreaction exist, the design of the system is not capable of detecting or transmitting the trigger.

1

A risk linked to the independence of the system form any other control outside from its design is that it might not be possible to turn off an autoreactive device, unless it has been designed for it. If, for instance, the autoreactive proprieties are embedded within a material, it cannot be prevented from reacting if the boundary conditions are the ones it is programmed to react to.

Physical limits of the system can be quite varied, ranging from the limits in operation of the technology itself to the context it is integrated in:

- The upper and lower limits of operation range of the autoreactive system—as sensibility to the reaction trigger, magnitude of forces that can be transmitted, etc.—are mostly defined by its design and the technology embedded in it.
- Wear and fatigue of the material or of any other mechanical parts.
- The integration of moving parts into other structures can give rise to unwanted side effects as vibrations, etc.

1.6 Conclusion

The strength and the weakness of technological development is that, as it develops at an exponential rate (Encyclopædia Britannica 2019), its consequences cannot be assessed over a longer timeframe. Ideally, a technology should become more resource-efficient the more it develops—but on the other hand, according to Gartner's hype cycle (Gartner 2020), the innovation potential of a specific technology tends to stabilize in time. Should initial energy-intensive and polluting versions of a technology still be brought forward, assuming that more sustainable versions will be developed in time? Then, if there is an upper limit to how sustainable a technology can get, can that be considered to be sustainable enough?

We need to move beyond the logic of "the less harm the better" and strive instead for truly regenerative concepts. We cannot afford anymore to develop technology that is an end to itself. Technology needs to become resource-efficient while serving a purpose.

Motion is one way to introduce adaptability in artificial systems. Different types of kinetic technologies have been reviewed in this chapter, with the outcome that the majority of these solutions search for more refined levels of responsivity rather than striving for optimization for a set purpose.

To fill this gap, a new concept of reactive system is here suggested as a technological evolution in terms of resource optimization. Autoreactive systems are defined and described as autonomously acting systems that use available latent energy sources from their surrounding environment to achieve, for a set purpose, kinetic change. Central to the purpose and the use of autoreactivity is the set context of operation which defines energy sources, timeframes of operation, technologies and hence also most of the limits to the system itself. Autoreactive systems are suggested in substitution or in integration to other technical systems, in the parts of the system that can be designed to be autonomously regulated.

Further aspects that are central to the initial design stages of an autoreactive system—geometry, mechanics, energy and materials—are further studied in separate chapters, in integration to the fundamental definitions and principles here explored.

Time and behaviour are essential dimensions to control in any artefact that we use or that surrounds us. As designers and inventors of our own artificial reality, we need to accept change as a fundamental condition and become increasingly active parts of these processes. We need to rethink the nature of our objects and environments, starting from shaping new ideas and concepts to transforming the cultural and ethical approach to these topics.

Our technological advancements transform the world, in many cases well beyond our initial aims, and in some not towards the better. This lays on us an enormous responsibility already in the first stages of conception and design of new solutions. A sound criticism towards an unleashed technological development is a growing necessity in a world where many efforts are spent towards progressing in the wake of the technological overshoot rather than spending the resources on truly progressive solutions. We need to ask ourselves critically what the products of our efforts are, what are the benefits and the costs we need to sustain.

References

Aelenei L, Brzezicki M, Knaack U, Luible A, Perino M, Wellershoff F (2015) Adaptive facade network—Europe. TU Delft Open for the COST Action 1403 Adaptive Façade Network, TU Delft, Delft. ISBN 978-94-6186-581-6

Beesley P (2010) Hylozoic ground, liminal responsive architecture. Riverside Architectural Press. ISBN 978-1-926724-02-7

Benyus J (2002) Biomimicry: innovation inspired by nature. Harper Perennial, New York

Calder A (2015) Hanging mobile. The Calder Foundation. Retrieved 10 Feb 2015 from ► http://www.calder.org/work/by-category/hanging-mobile

Encyclopædia Britannica (2019) Moore's law. Encyclopædia Britannica, inc. Retrieved 03 Apr 2020 from ► https://www.britannica.com/technology/Moores-law

Fox MA, Kemp M (2009) Interactive architecture. Princeton Architectural Press, New York

Gartner (2020) Gartner hype cycle, interpreting technology hype. Gartner, Inc. Retrieved on 03 Apr 2020 from ► https://www.gartner.com/en/research/methodologies/gartner-hype-cycle

Gibson DG, Glass JI, Lartigue C, Noskov VN, Chuang R-Y, Algire MA, Benders GA, Montague MG, Ma L, Moodie MM, Merryman C, Vashee S, Krishnakumar R, Assad-Garcia N, Andrews-Pfannkoch C, Denisova EA, Young L, Qi Z-Q, Segall-Shapiro TH, Calvey CH, Parmar PP, Hutchison CA III, Smith HO, Venter JC (2010) Creation of a bacterial cell controlled by a chemically synthesized genome. Science 329(5987):52–56. ► https://doi.org/10.1126/science.1190719

Gruber P (2011) Biomimetics in architecture, architecture of life and buildings. Springer-Verlag, Wien. ISBN 978-3-7091-0331-9

Harris J, Wiggington M (2002) Intelligent skins. Elsevier LTD, Oxford. ISBN 978-0750648479

Janocha H (ed) (2007) Adaptronics and smart structures: basics, materials, design, and applications, 2nd edn. Springer-Verlag, Berlin. ISBN 978-3-540-71965-6

Jormakka K (2002) Flying dutchmen: Motion in architecture, Birkhauser, Basel. ISBN 978-3764 366391

Kahn N (2006) Wind silos, International Trade Center, Charlotte, North Carolina. Ned Kahn Studios. Retrieved on 5 Apr 2020 from ► http://nedkahn.com/portfolio/wind-silos

May J (2010) Handmade houses & other buildings, the world of vernacular architecture, 1st edn. Thames & Hudson, London

Moloney J (2011) Designing kinetics for architectural facades: state change. Routledge, Abingdon. ISBN 978-0-415-61034-6

Oosterhuis K (2012) Hyperbody forward to basics. Hyperbody 2012. Retrieved 11 Nov 2014 from ▶ http://www.hyperbody.nl/fileadmin/pdf/HyperbodyMasterBooklet_2012.pdf

Oxford English Dictionary (2020) Functional. Lexico, Oxford English Dictionary. Retrieved on 30 Mar 2020 from ▶ https://www.lexico.com/en/definition/functional

Persiani SGL (2018) Biomimetics of motion, nature-inspired parameters and schemes for kinetic design. Springer International Publishing, Cham. ▶ https://doi.org/10.1007/978-3-319-93079-4

Persiani SGL, Battisti A, Wolf T (2015) Autoreactive architectural facades—discussing unpowered kinetic building skins and the method of evolutionary optimization. In: Proceedings of the 11th conference on adaptive building skins, Bern, 10–11 Oct 2015. ISBN 978-3-98120539-8

Persiani SGL, Molter PL, Aresta C, Klein T (2016) Mapping of environmental interaction and adaptive materials for the autoreactive potential of building skins. In: Proceedings of the 41st IAHS world congress sustainability and innovation for the future, Algarve, 13–16 Sept 2016, p 302. ISBN 978-989-98949-4-5

Ritter A (2007) Smart materials in architecture, interior architecture and design. Birkhaüser, Basel. ISBN 13: 978-3-7643-7327-6

Zahner (2009) Zahner, works, Simons Center for Geometry & Physics, Stony Brook. Retrieved 5 Apr 2020 from ▶ https://www.azahner.com/works/stony-brook

Zuk W, Clark RH (1970) Kinetic architecture. Van Nostrand Reinhold, New York

Morphosis of Autoreaction

Contents

S. G. L. Persiani, *Design of Autoreaction*,
Design Science and Innovation, https://doi.org/10.1007/978-981-15-6178-8_2

The morphology or the shape of a moving system is much more complex to represent, to understand and to design than that of a static object as it changes in time, in space and sometimes also in function.

In this chapter, the aspects concerning the morphosis, or the "shaping", of the autoreactive system are discussed, from the spatial framework (the morphospace) within which the change happens, the geometry (morphology), the process of shape change (the morphosis), to the capacity of the shape and the motion features to add a new dimension and unique identity to the artefact.

2.1 Morphospace

The morphospace is, in evolutionary biology, an imaginary multi-dimensional space where different organisms can be compared through their different morphological parameters, represented on different axes (Valentine 2004). In a technical context, the morphospace is rather seen as the special trajectories that a body's shape follows through space and time (Steadman and Mitchell 2010). The morphospace, therefore, defines all the positions that the object will assume acting along specific planes and paths from the starting morphology to its final morphology.

Defining and describing the parameters of a body in motion within that morphospace is essential, not only to understand and shape the morphology of an autoreactive system, but also to control its movement.

2.1.1 Frame of Reference

To describe the geometrical features of an element, a frame of reference is needed as the body can have different shape and kinetic proprieties depending on how it is oriented. A shared coding system is also necessary to describe relationships of parts within the system itself, and ultimately to represent the movement in the best way.

The virtual space used to represent a movement often resembles the body's own morphology, for an intuitive and straightforward description and understanding of the movement. In music, the staff is divided symmetrically according to the mode of the execution (a staff for the right hand and one for the left hand for the piano), while dance notation is characterized by symmetrical features which recall the structure of the body which performs the movement (Maletic 1987).

In the context of kinetic man-made artefacts, planar geometric right-angle projections are typically used, as in the case when representing static elements. Descriptive geometry allows to represent three-dimensional objects in two dimensions by creating planar geometric right-angle projections of the object within a Cartesian coordinate system, with plan, elevation and sections as outcome. This method of representation, however, presents some major flaws when aiming to fully describe a movement.

First of all, depending on the geometry of the object and the movement performed, variations in the virtual space might be more suitable: in the case of a kinetic five-pointed star element, a pentagrammic representation might be more suitable than a two-dimensional Cartesian coordinate system.

Secondly, the same movement can appear very differently depending on the point of view it is observed from. To be fully able to describe a movement within a system, and the relationship between the moving parts and the static parts within that system, it is important to use a single frame of reference to distinguish between planes of projection of the movement.

To fully describe the complexity of the movements relatively to each other, inspiration is taken from the anatomical planes—an abstract concept used to describe human and animal body anatomy in a coordinate system. The anatomical planes are four imaginary planes that pass through a body, following its geometry, that are then used to describe anatomy, location and movements (Ogele 2013). In the context of autoreactive systems, an adaptation of this same concept is made to describe the geometrical features and their movements. The virtual planes are (▫ Fig. 2.1):

- Median plane (Anat. sagittal plane), the single vertical plane that passes through the main axis of symmetry of the body, dividing it into right and left halves. In bodies with more than one axis of symmetry, the median plane is the first reference to be set, as all other planes then refer to it;
- Paramedian planes (Anat. parasagittal planes) will be used to refer to all other planes that are parallel to the median plane, but do not divide the body on its main axis of symmetry;
- Frontal planes (Anat. coronal planes) are all the planes at right angles to the median plane passing through the body, dividing it into front and back portions. In those bodies, where a specific axis of symmetry identifies front and back, the frontal plane passing through that axis will be referred to as mid-frontal;

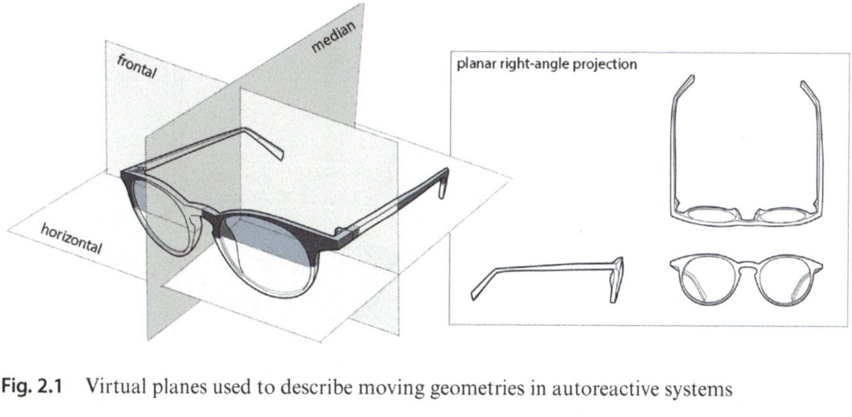

▫ **Fig. 2.1** Virtual planes used to describe moving geometries in autoreactive systems

- Horizontal planes (Anat. transverse planes) are all the planes at right angles to the median and the frontal plane passing through the body, dividing it into top and bottom. In those bodies, where a specific axis of symmetry identifies top and bottom, the horizontal plane passing through that axis will be referred to as mid-horizontal.

Additional terms used to describe the virtual planes above more in detail are:
- View, when the plane does not intersect with the body;
- Section, when the plane intersects with the body.

As can be noticed looking at many examples, the frontal view is one of the most commonly used views to represent the geometry of a man-made artefact. To represent a specific movement, however, the views of the planes where the action occurs are usually represented: human running, for instance, occurs in the median and the horizontal planes, while jumping occurs in the median and frontal planes.

2.1.2 Keeping Track of Movements

Reversible moving systems, which are the object of focus in this context, can generally be described and controlled by defining at least three of its moving stages (Fig. 2.2):
- The starting position (p_0), which is the position or shape taken as reference before the system is set into motion. It is generally identified as a resting position for the system, such as a closed door which perfectly fits its frame;
- The limit position (p_{lim}), which is usually the position furthest away from the starting position, where the system needs to be returned to, such as a door fully opened to its widest point at 180 degrees;

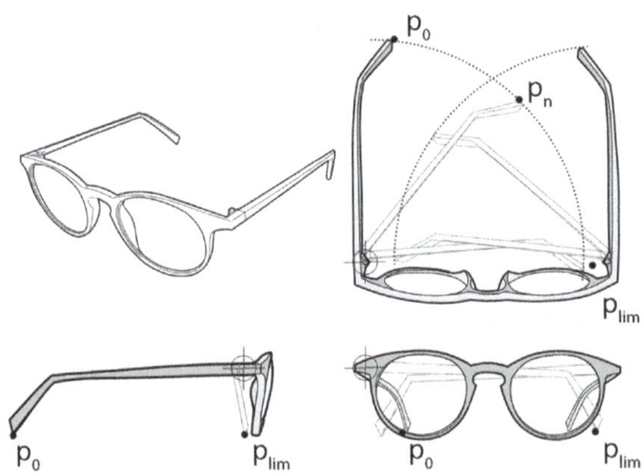

Fig. 2.2 Example of the rigid movement in an object using a planar geometric right-angle projection showing the movement trajectories, the point of attachment to the main body and the free end with three moving stages p_0, p_n and p_{lim}

- One or more intermediate positions (p_n), helping to identify the nature of the movement in between the starting and the limit position. Many movements move from p_0 to p_{lim} and then reverse the same identical movement to return to p_0, while others proceed with a new movement to return to p_0. In the second case, more than one intermediate position might be necessary to define the movement.

2.1.2.1 Transition Between Morphologies

When representing moving three-dimensional elements, the projection where the change of shape is the most representative is generally chosen and the different shapes undertaken at regular time intervals are overlapped. This type of representation mostly focuses on the geometric aspects within the movement, aiming to display their evolution rather than describing the parameters of the movement itself.

2.1.2.2 Trajectories

The most common way of describing a geometry in motion is to identify in each movable part, a point of attachment to the main body (or to other body parts) and at least one free end. Theoretically, the maximum movement is performed at the free end while no movement is performed at the point of attachment. The free end is moving relatively to its point of attachment in a specific direction described by a direction symbol. Key points on the moving part can also be identified and followed by drawing the trajectory of these specific points (Fig. 2.2).

Further variables to take into account, such as folding and bending points, variating degrees of contraction and elongation, etc. can be described with appropriately defined symbols.

2.1.2.3 Centre of Gravity

Keeping track of how the centre of gravity (*G*) changes during the transition from one geometry to another in the body is a key issue to control the mobility of the body, as well as its balance, and kinetic potential (these aspects are further discussed in ► Chap. 3) (Fig. 2.3).

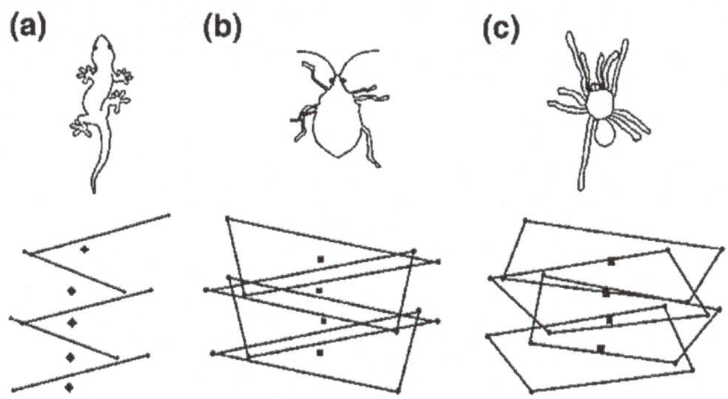

Fig. 2.3 Representation of how the centre of gravity (*G*) shifts during the slow walking gait of **a** a gecko, **b** a stinkbug and **c** a spider. The dots highlight the shifting of (*G*), while the lines indicate feet-holding positions. From Ji et al. (2014)

Generally, the possible phases that the centre of gravity can undergo are five in number:

1. *G* does not move, which means that either the whole body remains static, or parts of the body move so as to neutralize each other;
2. *G* is shifted, which means that its location is adjusted within the space occupied by the body so as to improve on a specific performance;
3. *G* is transferred, which relates to a displacement of the body in space;
4. *G* can be identified but no part of the body actually carries the weight, which happens during jumps or flight phases;
5. *G* rotates.

2.1.2.4 Timeframe

The timeframe is a central aspect in the description of a motion. It is therefore surprising how few time-related parameters are found in technical representations of kinetic systems.

Ideally, the virtual space representing the movement is subdivided in sections representing a time unit. The bars that set the basic time unit use a constant amount of space or can be flexible depending on the amount of information that needs to be indicated in between two time units. This way, the movements described inside that frame are given a specific duration (■ Fig. 2.4).

Within the defined timeframes representing successive movements, patterns of one or more factors can be identified as repeating themselves (geometries, movements, sounds, etc.). Controlling and shaping these patterns is a powerful tool to create unity within a work, but also to make the work dynamic by providing recognizable milestones for the observer to measure the change with. This opens up to completely different parameters that characterize the movement, the perception of the scene and conveying of emotions.

The pace of a movement strongly defines a personality and character in objects by conveying sensations and emotions to the observer. Speed gives sensations of excitement, stress, effectiveness while slower movements provide feelings of control, relaxation and encourage observation. This is why some types of movements can also be a bad or a good choice in terms of user experience. For instance, the fast and aggressive movements of a car are purposefully designed

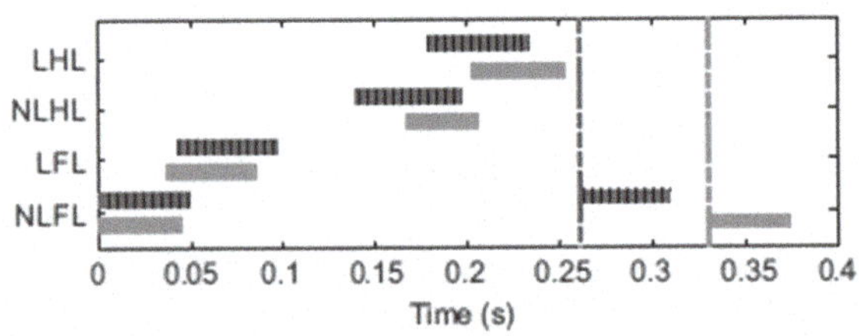

■ **Fig. 2.4** Biomechanical example of a movement described within a specific timeframe: comparison of the footfall patterns in a greyhound (dark grey) and a cheetah (light grey)—non-lead forelimb (NLFL), lead forelimb (LFL), non-lead hindlimb (NLHL) and lead hindlimb (LHL). Modified from Hudson et al. (2012)

to give the impression of dealing with an organism of higher intelligence such as an animal, while slower movements reminding more of a plant's slow reactions might fit a building skin better.

Similarly, the transition between phases, movements or gaits modulates the dynamic flow, accompanying the movement into a new pattern ensuring a smooth and effective transition between states. This happens during the slow in and slow out: in technology as in nature, systems need to go through different gaits, or gears, to allow the system to efficiently slow down before reaching a complete stop or speed up reaching full speed. The design of these stages is important from an energetic point of view, but it also enhances the feeling of the observer for the movement.

2.1.2.5 Relationship Between Parts

The relationships between movements within specific timeframes are comparatively used less in design and engineering and are much more developed in other fields such as biomechanics, music or dance, where these parameters have a recognized fundamental importance. It is however through the design and control of these kinds of relationships that a great number of characteristics of a movement are controlled, in terms of function, energy efficiency and aesthetics alike.

The arrangement and frequency of repetition of movement patterns in time, whether it is referred to as rhythm (music) or gait (biomechanics), and characterizes a great number of fundamental functions that are an intrinsic part of moving patterns (◘ Fig. 2.5).

A gait (biomechanics) is defined by the timing and frequency with which an animal performs a specific movement (footfall) pattern: such as jumping, crawling, walking, skipping or running. Changing gait, as a machine changes gear, is a way for a system to adapt to different speeds, terrain, manoeuvres, etc. and guarantee energetic efficiency. Specific gaits are used for specific speeds: as fast walking becomes strenuous for the organism we intuitively switch to running where a different oscillating system allows a more efficient energetic return at a lower energy input (Persiani 2018).

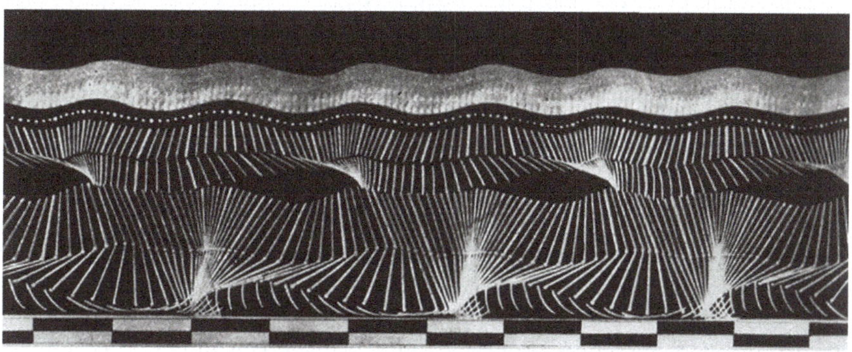

◘ **Fig. 2.5** Chronophotographic sequence of the rhythmic pendulum-like walking pattern of a man (from Marey 1884). The sequential movement of the head, the upper and lower limbs are identified through markings on the body

We have evolved an innate sensitivity and attraction to rhythms. Falling into a specific rhythm or beat is an exciting active experience for humans, which often closely relates to beauty, wherein a variation is also essential. One manifestation of this tendency is entrainment, a sort of magnetic effect with which one rhythm creates an attraction upon another rhythm, influencing its frequency and drawing the phases to change timing until they couple to the same frequency (Bush and Clarac 1985). This happens as people walking next to each other fall into the same walking frequency, or when music influences coordinated motion, such as during treadmill walking, running and dancing (Leman et al. 2013).

Among the many variations that can be identified within patterns of action and can also help the description of the effect(s) achieved, two examples are:

- Pitch (much used in music) which is a subjective sensation of reaching a high or low peak in a succession of actions. In a physical movement, this can be reflected by the maximum or minimum expression of a movement. In all contexts, the design of the way (modus) of reaching the climax is of extreme importance for maximizing the aesthetic and energy expression of the movement;
- Climax, which individuates the building up of energy in the pattern (through acceleration, succession of size change, sound, accumulation, etc.) to reach the highest or a high point in the overall movement. In energy terms, this kind of pattern will be connected to a specific kinetic design, while in aesthetic terms, it reflects in a building up of expectations from the part of the observer which is connected to the comprehension of the overall pattern and anticipation of the rising of defined parameters.

2.2 Morphology

Motion strongly depends upon the geometrical features of the elements it is made of. Factors such as inertia, weight, energy absorption and balance depend on geometry. Therefore, the choice of fitting morphological features, as well as an integrated development of morphological, mechanical and energetic factors is of fundamental importance to design systems with an optimized capacity to move.

In nature, where kinetic systems have developed and been selected over millions of generations and trials, morphology, movement type and movement control have co-evolved exploiting the morphological features in energetic and mechanical contexts to the utmost (Pfeifer 2000). If we assume that each one of these combinations of physical, mechanical and geometrical features have adapted to the specific physical conditions it operates in at a maximum output and a minimum energetic input, then the observation of the combination of these features is enormous evolutionary advantage in technical terms (Persiani et al. 2016).

In the following section, fundamental parameters defining morphological typologies and the dynamic potentials or drawbacks that are the consequence of using these features are discussed.

2.2.1 Symmetry

Engineering structures as well as most organisms in nature are often built with some degree of symmetrical or geometrically organized body plan.

Symmetrical features not only simplify design and construction processes, as a way of forming patterns and connections through the repetition of simple sets of information (Gruber 2011; Persiani 2018), but often also represent an optimum in terms of efficiency (Várkonyi and Domokos 2007).

Setting aside the graphic or aesthetic use of symmetric and asymmetric features, the design of the body shape strongly affects the kinetic characters of a system. Correlation between morphological symmetry and locomotive efficiency has been indirectly demonstrated in biology and technology, as systems with a higher degree of symmetry tend to exhibit greater locomotive efficiency than organisms with higher asymmetry (Persiani 2018; Valentine 2004).

There are mainly three types of symmetry—reflection, rotation and translation symmetry—and a number of possible combinations of these.

2.2.1.1 Reflection Symmetry

Reflection symmetry is also known as linear or bilateral symmetry. A figure has reflection symmetry if it can be divided into two identical, mirrored halves by a single line identifying the axis or the plane of symmetry.

Linear symmetry suggests an almost monodimensional mirroring along the median plane, and partially along the horizontal plane, reproducing a sequence of similar elements along the central body axis (◘ Fig. 2.6).

In biological terms, *bilateral symmetry* is so important that in Linnaean taxonomy (the science naming, defining and classifying groups of biological organisms based on their shared characteristics), organisms displaying "bilateral" symmetry are classified into one same monophyletic group—which include most animal species apart from sponges, medusae, polyps, comb jellies and basal multi-cellular organisms (Maiorana and Van Valen 2020).

In this context, the axis of reflection is generally identified with the median plane (see ► Sect. 2.1.1. Frame of reference) (◘ Fig. 2.7).

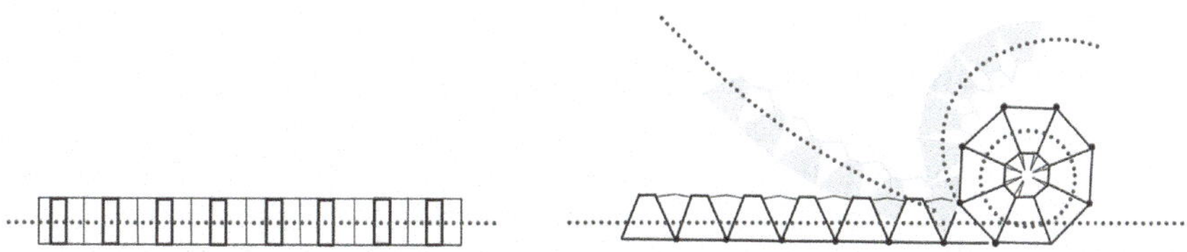

◘ **Fig. 2.6** Linear symmetry in the Paddington Basin Rolling Bridge, top and side view. The bridge would display perfect linear symmetry if it was not for its kinetic operation, which enables the bridge to roll up only on one side of the canal (see also case study in ► Chap. 6)

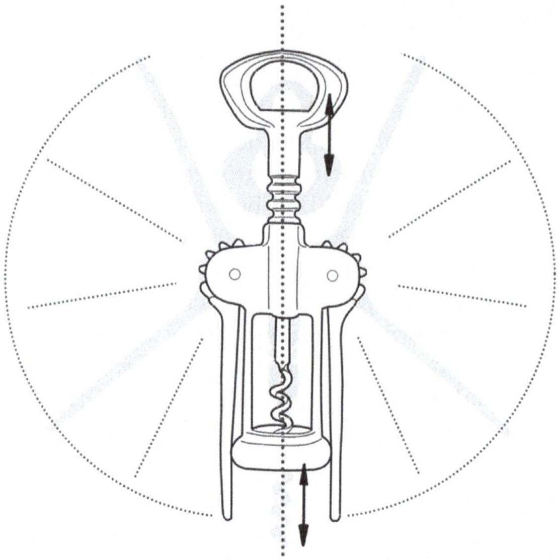

Fig. 2.7 Bilateral symmetry in a corkscrew

Typical kinetic characteristics of systems displaying reflection symmetry are:

- Directionality of motion. The symmetry axis often identifies the preferred direction of motion, which can also be further subdivided into a “forwards” and a “backwards” direction, meaning that motion happens mainly in one of the two directions highlighted by the axis, the opposite direction (when enabled) being used mainly to enable manoeuvres;
- Differentiation of functional body parts. As the parts allotted for locomotion can specialize on locomotion in one prevalent direction, other body parts are freed for other uses;
- Fluid dynamic proprieties. As anything that moves around on our planet needs to move through at least one medium, and mostly a fluid such as air or water, the body shape is also important in terms of fluid dynamics. Fluid dynamic shapes are in fact mostly directional (with a front and a rear) and exhibit reflection symmetry at least in one dimension.

2.2.1.2 Rotation Symmetry

Rotation symmetry is developed multi-laterally and equidistantly starting from a central point. It can be further subdivided into radial and spherical symmetries.

Radially symmetrical shapes are developed bi-dimensionally, circularly or along axes (tri-, tetra-, penta-, hexa-, octa-merism) (Fig. 2.8). In some cases, radial symmetry does not present any mirroring axis, but is rather the product of the rotation of an asymmetrical shape. *Asymmetrical radial* shapes are very common in mechanics, and especially in gearing, where the asymmetrical proprieties of the teeth confer different proprieties to the mechanism depending on the direction of rotation of the gear (see also ▶ Chap. 3).

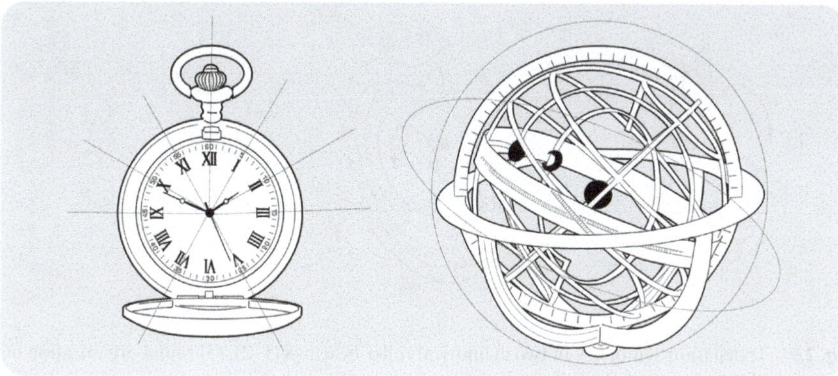

Fig. 2.8 Radial symmetry in a pocket watch (left) and a close to spherical symmetry in an armillary sphere

Spherical symmetries are, on the other hand, three-dimensional, meaning that the system can be cut into two identical halves through any cut that runs through the geometrical centre (Fig. 2.8).

Typical characteristics of systems displaying rotation symmetry are:

- Multi-directionality of motion. Ideally, systems with rotation symmetry are able to move, function or operate in more than one direction in the plane where they exhibit rotation symmetry;
- Multi-functional body parts. As all radial body parts resemble each other, the functional specialization of body parts is no longer possible, and several functions need to cohabit and be managed by each segment;
- Dynamic stability in the planes that do not show any radial symmetry; dynamic rotational potential in the planes that show radial symmetry, with a centred centre of gravity.

Depending on how extreme the rotation symmetry in the body is, and whether it presents any reflectional symmetrical elements, the kinetic characters that can be achieved are very varied. An example is animals with more than four legs which, although properly bilateral, exhibit radial kinetic proprieties as the ability to move sideways as well as in the direction of their heads (Biewener 2007; Cruse and Graham 1985).

2.2.1.3 Translation Symmetry

In *translation symmetry*, a geometry is reproduced in another position while maintaining its general or exact orientation. The intervals between the geometries do not have to be equal in order to maintain the translational symmetry, but generally bind the shapes together by respecting some kind of proportion. Translational symmetry is very important in nature and artefacts alike, as it is used for multiplication principles such as segmentation and pattern-creation.

Segmentation often combines the translation symmetry with a linear symmetrical development: more or less identical subunits are used to form a structure, as

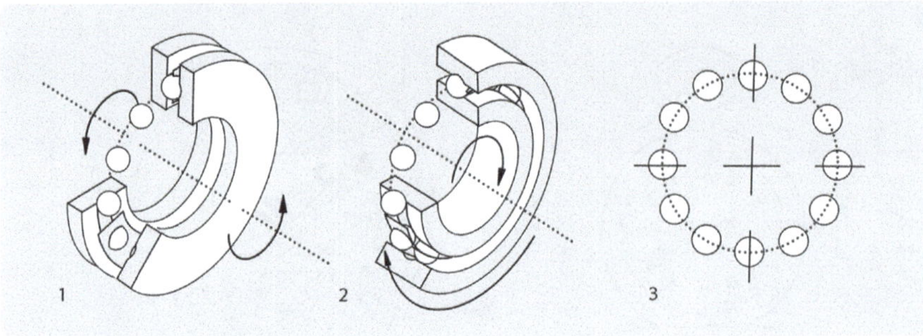

Fig. 2.9 Translation symmetry in two cylindrical roller bearings (1–2); (3) radial organization of the segmented translation of the roller spheres

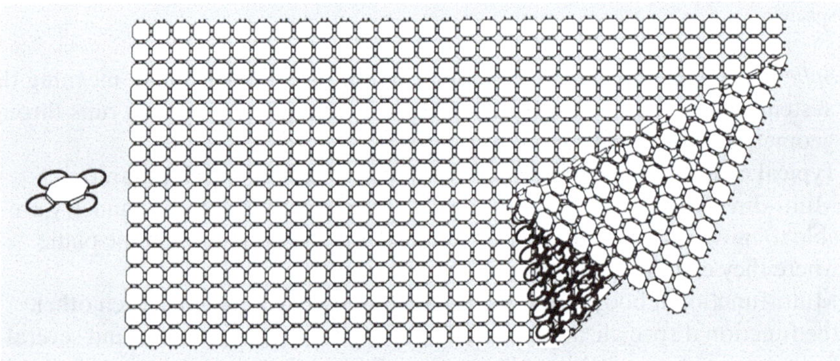

Fig. 2.10 Example of pattern-creation of a metallic fabric out of a simple element with four connections

in the bones of a spinal cord. A great number of mechanical systems, and in particular chain systems, are created using this technique (Fig. 2.9).

In *pattern-creation*, the translation happens in a bi-dimensional or three-dimensional dimension rather than in a monodimensional one. What is mostly interesting, from a dynamic perspective, in these patterns is—more than the overall geometry of the system itself, the connections that exist between the reproduced geometries and the intervals that binds each part together with its neighbour (Fig. 2.10).

Typical characteristics of systems displaying translation symmetry are:

- Directionality of motion mostly happens in the planes orthogonal to the translation plane, as the latter locks the single parts together to some extent;
- Single elements tend to be monofunctional, but the system has a potential for creating complexity and emergence of behaviour on a larger scale.

2.2.1.4 Symmetry Combinations

Systems combining two or three base-symmetries also exhibit a mix of the dynamic characters of these.

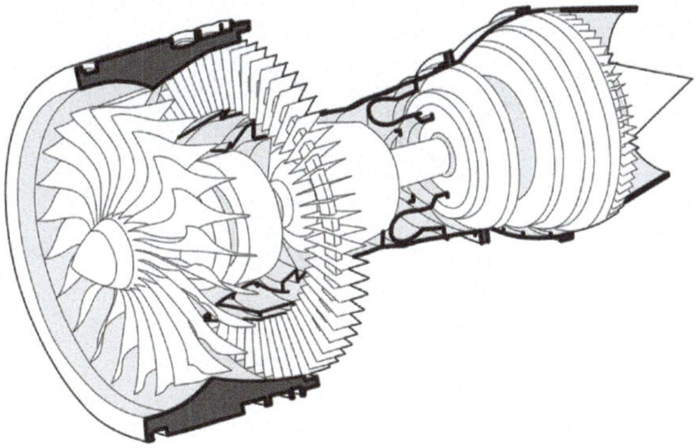

Fig. 2.11 Bi-radial symmetry in the design of a sectioned turbine

The *bi-radial symmetry* is a particular combination of a radial and a bilateral symmetry that can be found in some natural systems such as comb jellies. In this case, the system exhibits reflection symmetry on the median plane and radial symmetry on the horizontal plane, through a succession of different elements (Fig. 2.11). Bi-radial symmetry is very common in mechanical systems, where gear trains and turbines can display this type of symmetry.

Spiral symmetries can be seen as a combination of the radial and the segmentation symmetries, where the geometry is constantly reproduced along a line, around a fixed central point/axis individuated outside the physical boundaries of the organism, at a continuously increasing or decreasing distance from the point. Natural systems using the spiral of Archimedes, the logarithmic spirals and Fibonacci sequences are quite common as this type of shape formation is a way to preserve energy and proportions of an organism during growth (Wolpert et al. 2011) (Fig. 2.12).

Helix symmetries combine the radial and linear symmetries by developing, turning around the axis while moving parallel to the axis. Double-helix symmetries add a rotated translation to the normal helix (Fig. 2.12).

Screw-like symmetries combine the spiral symmetry and the linear symmetry to obtain a helix where the distance from the centre of rotation increases or decreases at a constant rate (Fig. 2.12).

2.2.1.5 Asymmetry

An asymmetric system presents no symmetric proprieties at all throughout its overall body structure, and in all three dimensions.

Total asymmetry is relatively rare, both in nature and in man-made things. Generally, systems with higher asymmetry present several handicaps among which:

- Difficulty of use (for tools). A totally asymmetric tool is not only more difficult for a user to grasp (while on the other hand, a tool's symmetrical features give clues about its possible function) and therefore also to use;

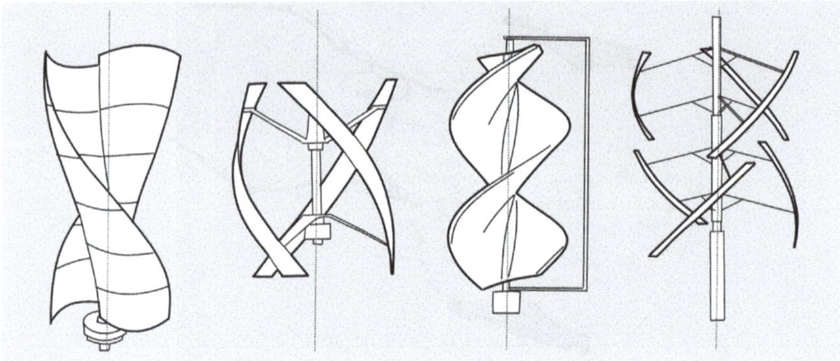

Fig. 2.12 Spiral, helical and screw symmetries in different designs for wind turbines

- In kinetic terms, highly asymmetric bodies tend to have erratic and deviating movements, therefore, requiring a higher energy expenditure to counterbalance and stabilize motion (Wolpert et al. 2011; Arthur 1997).

Partial asymmetry, on the other hand, meaning small deviations from the perfect geometrical symmetry, is extremely common in nature—from opposable thumbs to the asymmetric placement of important organs such as the heart. Generally, these exceptions are connected to a functional propriety or specialization of that body part (Valentine 2004).

Partial asymmetry is much less common in engineering problems, where asymmetrical proprieties are often correctly interpreted as un-optimal and imperfect. The use of small localized geometrical perturbations to improve on the response of engineering structures has, however, been theorized and suggested (Várkonyi and Domokos 2007).

2.2.2 Scale

Similar geometries are often used across very different scales, however, the size of moving parts in relation to their function and to their means of operation have a determining effect on the complexity of the kinetic system in terms of interaction with the surrounding medium, structural sizing, energetic and locomotive parameters. These geometries are here called *sibling geometries*, meaning they exhibit the same geometry but at different scales in size.

In fact, as the design of a mechanical system (natural or artificial) is copied and reproduced at a different scale, its dimensions and proprieties cannot be uniformly sized up or down: there are structural and functional consequences to the changes made to the linear, areal and volumetric dimensions. As these are scaled at different rates, all processes from structural design to energy balance are strongly affected by these parameters.

When the same geometry is applied to dimensionally different scales, three basic design parameters must co-evolve together with the geometry to ensure a successful kinetic outcome (Schmidt-Nielsen 2004):

- Dimensions, as structures grow progressively thicker and bulkier as the size grows;
- Materials, which typically need to be progressively replaced, from lighter and flexible materials at small scales to more rigid alternatives at larger scales;
- Design, changing the structural behaviour of the system, such as from compression to tension elements.

These three aspects need to be integrated so as to solve different issues at different scales while ensuring the system's efficient operation, according to the end purpose. Three of the most recurrent size-related changes in conditions are here further discussed.

2.2.2.1 Medium of Motion

The medium of locomotion is defined as the type of environment that the system moves through or on: aquatic (in or on water), terrestrial (on ground or other surfaces), fossorial (underground) or aerial (in the air).

The nature and proprieties of the medium play an important role in the design of the movement, which it influences it in different ways on different scales. This, for instance, becomes very obvious on very small scales, such as in microstructures of one centimetre or smaller. Very small autoreactive systems moving in fluids (air, water or other fluids) at low Reynolds numbers, need to be designed considering that viscous forces dominate rather than turbulent ones. As turbulence is absent, the aerodynamic or fluid dynamic shape of the system becomes largely irrelevant and even counterproductive as the ability to move through the fluid is more bound to the body's capacity to reduce friction drag (Biewener 2007). As any symmetrical movements would cancel each other out, these systems must rely on more asymmetrical kinetics to achieve propulsion.

It is, therefore, important in these contexts to distinguish between systems where a perfect kinetic performance is not vital and success can be achieved, although the system does not unfold at its best, and systems where the good kinetic performance is inseparable from the task to be performed. The first case is, for instance, a system where the reaction is mainly aimed at unlocking a specific potential—as in remotely activated biomedical devices performing drug delivery (Genchi et al. 2017). Here, the dynamics of the kinetic reaction are secondary to the timing of the reaction: what is important is that the content is delivered at the right moment. The second case is when the dynamics of the movement itself are central to the task to be performed—as in self-folding mobile microrobots where not only the folding sequence and coordination is important to achieve the intended shape but also to enable the robot to move around and carry loads (Hardesty 2015). In this latter case, the medium becomes a main parameter in the device's design.

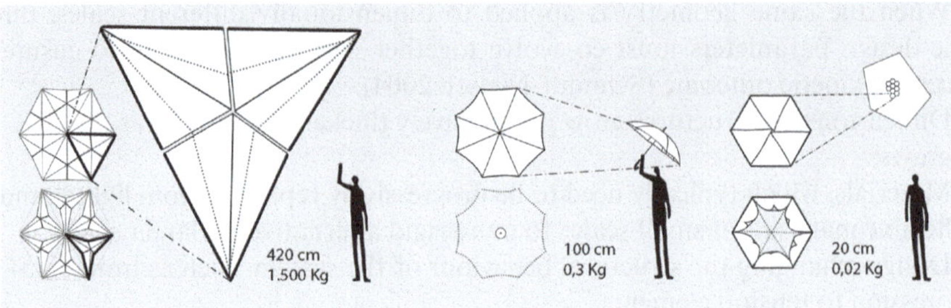

Fig. 2.13 Examples of three openable systems and their differences in design, operation, size and weight. (1) The sunscreen elements in the Al-Bahr Tower in Abu Dhabi. Adapted from Attia (2016); (2) an umbrella; (3) a panel of the HygroSkin pavilion (ICD 2013)

2.2.2.2 Weight of the System

An aspect that radically limits the design of kinetic system is the size and the weight of the system itself. As the size of the design grows, large-scale constructions typically present bulkier structural frameworks that weigh more and take up proportionally larger areas. This affects the body's inertia (see ▶ Sect. 3.1.4. Principle of Inertia for the definition), the physical propriety of all bodies to resist changes in their state of motion, which, becoming exponentially high as the size and the weight of the system grows, affects its capacity to move and be handled.

It is, therefore, not surprising that dynamic systems are rarely achieved on very big scales at all—unless strictly necessary. The dynamic proprieties are rather integrated by introducing smaller kinetic parts that consume in proportion less energy and are easier to handle and maintain (Fig. 2.13).

As anticipated, there are some context-specific cases where the size of the moving parts becomes an advantage rather than a drawback. This is, for instance, the case of wind turbines, where the amount of energy that can be extracted at a given wind speed is proportional to the size of the rotor—more exactly, to the cube of the wind velocity and to the square of the rotor radius (Gipe 2009). This means that a small increase in rotor diameter significantly increases the power generation of turbines. In this case, the manoeuvrability of the kinetic system is secondary to the speed of rotation and hence of the power output of the turbine.

2.2.2.3 Manoeuvrability Versus Speed

The manoeuvrability of a system—or in other words how easily it allows to be moved and directed—depends on the previously mentioned inertia of a body. This biomechanical principle can also be seen in natural systems, where, for instance, a bird's wing shape influences their patterns of flight. Long and narrow wings allow stable long-distance flights with little energy expenditure while short and stubby wings require a more energetic flight mode but allow very quick changes in direction (Biewener 2007).

The most movable systems are those that can make sharp and sudden changes of direction, while being characterized by low centres of mass, narrow and tendentially more varied shapes. As the weight is not too much of an issue in kinetic systems with a contained size, a bigger variation in shape can be achieved without significant losses in terms of energy efficiency.

The narrowness of the shape, on the other hand, although it reduces the body's inertia, therefore, increasing manoeuvrability has important consequences on the speed of the motion. Since stride length (or the length of a lever) is a clue factor in speed, systems with small body shapes have in fact limited possibilities to achieve speed compared to larger body structures. Low speed is, therefore, the other side of the coin of manoeuvrability, and vice versa: objects travelling at fast speeds increase their inertial mass, resisting change and becoming less manoeuvrable.

2.3 Morphosis

The morphosis is a concept that describes the change of a shape in time and space—it therefore no longer describes one shape, but a whole process of transformation from one shape to another. While in biological terms, the morphosis involves a physical formation or development through an irreversible additive process, in this context only reversible changes are taken into consideration.

In morphological terms, reversible mechanisms (mechanisms that can be returned to their initial shape before the action) are most often connected to specific geometries which accommodate and embed an intrinsic potential for change (Persiani 2018).

Overlapping is a way to organize, condense and protect a set of similar or identical geometrical shapes by allowing them to slide over one another until they take up the smallest surface possible. Overlapping geometries are often accompanied by a hinge system or similar, which allows to keep the shapes moving parallelly to each other (◘ Fig. 2.14).

The main advantages for which this type of motion geometry is used are:

- Chain reaction. In some cases, the shapes have a defined range of motion relatively to one another, resulting in the movement of the first shape to drive the sliding in all the successive parts;
- Fast deployment. The geometry is particularly high-efficient from an organizational and technical point of view, allowing the element to deploy in very short time lapses, without any risk of being jammed.
- Stability. Overlapping single areas can also be a way to effectively introduce a system-based surface stability, as in the case of a hand fan or in birds' wings.

Folding subdivides a broader surface in two or more areas (the faces) linked to one another through common bending edges (the folds) that can behave like rotational hinges or localized bending regions during folding (◘ Fig. 2.15).

The main advantages for which this type of motion geometry is used are:

- Storage. As the faces are made to rotate relatively to one another following the folds, and successively packed upon one another following pre-determined folding schemes, the initial surface can be compactly and efficiently stored

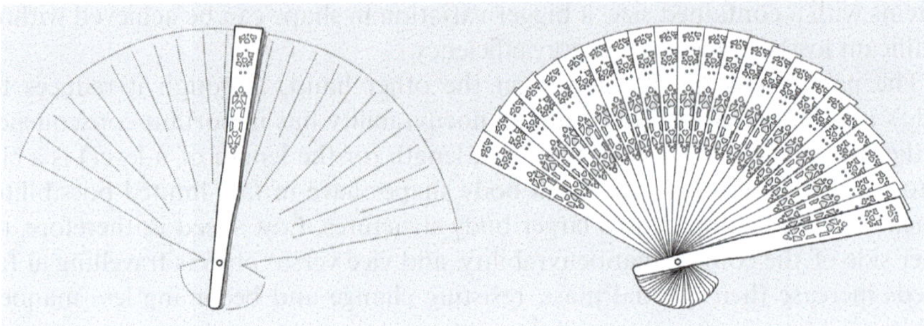

Fig. 2.14 Overlapping geometry in a wooden hand fan

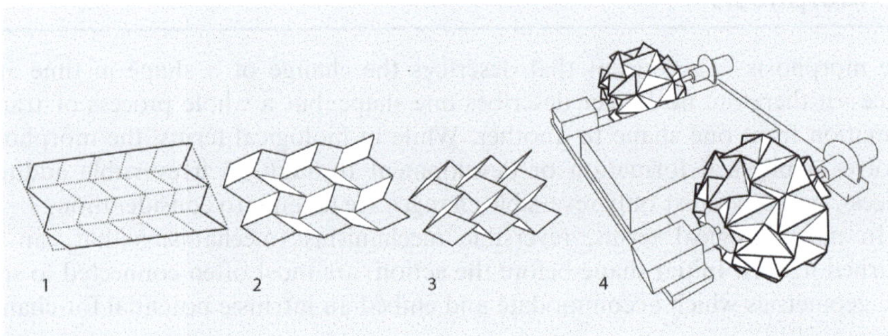

Fig. 2.15 Example of a Miura folding sequence (1–3); (4) application of a folded sheet into a wheel structure for a mini robot. Adapted from Lee et al. (2013). Thanks to the folding geometry, the size of the wheels can be manipulated through actuated folding and unfolding, allowing the robot to move in smaller spaces if needed

before the folding sequence is inverted and the surface deployed anew. These characters make folding structures among the most researched geometries in the context of self-deployable systems in engineering, allowing to attain a theoretically endless number of possible end-shapes using conventional origami folding techniques (Peraza Hernandez et al. 2019).

- Structural proprieties. A second typical use that is made of folding geometries derives from the added structural proprieties that can be embedded within a light and thin surface by introducing folding: the capacity of changing the structural stability, rigidity and capacity to resist to mechanical loads.

Bending is closely related to the folding principle, with the exception that the folding axis is found outside the physical boundaries of the organism, resulting in a more or less evident curving of the body (Fig. 2.16).

The main advantages for which this type of motion geometry is used are:

- Structural flexibility. This is a strategy that is often used in elastic systems to gather up kinetic energy within the geometrical structure. It is also a useful and simple movement in uniform body structures made of materials or substances that cannot integrate a hinge or folding point for different reasons;

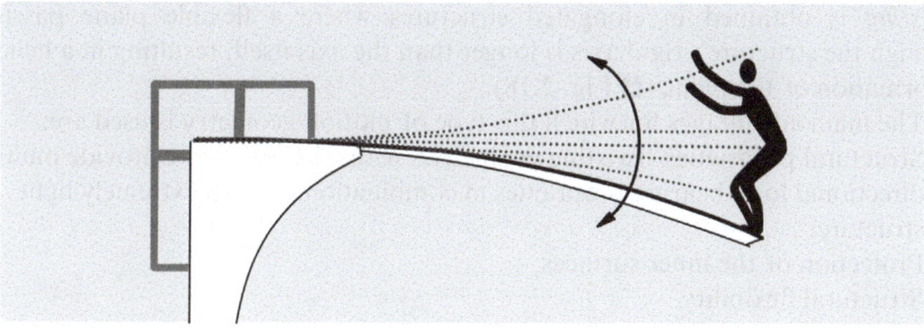

Fig. 2.16 Bending of a springboard

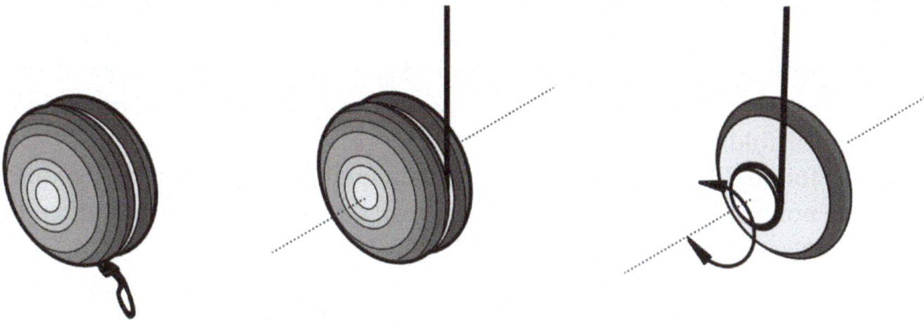

Fig. 2.17 Rolling of a wire around the axis of a *yo-yo*

- Dynamic flexibility. In some systems using bending geometries, the areas where the bending geometrically occurs are not always pre-determined, enabling the system itself to adaptively change the position and the magnitude of the bending itself.

Rolling is a way to organize large surfaces and slender structures, obtaining a compact shape of circular section (Fig. 2.17).

The main advantages for which this type of motion geometry is used are:

- Structural rigidity. Rolled-up shapes have a structural function, using the multi-directional load-bearing proprieties of the circular shape allowing to move large surfaces into their final position without deforming the surface during motion, and only successively deploying it—as in the growth of many leaf structures;
- Fast deployment. The geometry is particularly high-efficient from an organizational and technical point of view allowing the element to deploy in very short time lapses, without any risk of being jammed;
- Protection. In the rolled-up shape, the area of the surface that is in direct contact with the exterior is minimal, making it therefore an optimal way to protect the inner parts.

Twisting is obtained in elongated structures where a flexible plane passing through the structure's rigid axis is longer than the axis itself, resulting in a helical deformation of the plane (■ Fig. 2.18).

The main advantages for which this type of motion geometry is used are:

- Structural proprieties. Twisting geometries with a helical shape provide multi-directional load-bearing proprieties in combination with an extremely light structure;
- Protection of the inner surfaces;
- Structural flexibility.

Buckled, undulated or *crumpled* shapes are a way to compact mono and bi-dimensional surfaces by shortening the prevalent dimensional extension of the element by occupying as much space as possible in the other dimensions (■ Fig. 2.19).

The main advantages for which this type of motion geometry is used are:

- No pre-defined geometry. The main typical characteristic of this type of shape is that no specific geometry or folding system needs to be pre-determined for this type of folding to be achieved;
- Flexible folding. By organizing a surface occupying a larger area, the geometry does not need to use folding lines that can damage the surface. Rather, the surface gets a little crumpled but avoids sharp bending edges.

Thinning extension and *thickening compression* are the two geometrical extremes of a constant volume subject to linear change in dimensions: contraction and decrease in one of its dimensions result in an increase in another dimension (■ Fig. 2.20).

The main advantages for which this type of motion geometry is used are:

- Wide array of deformations. By controlling the various dimensions and parts of the body, many possible geometric deformations can be achieved, locally and in the overall system;

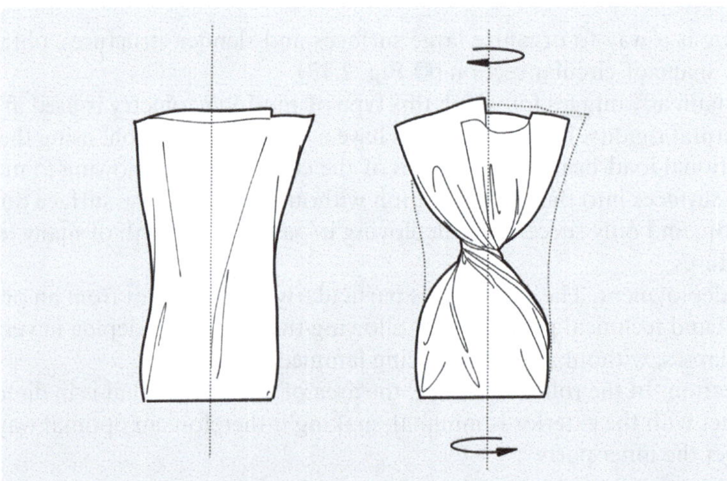

■ **Fig. 2.18** Twisting in the paper wrapping a candy

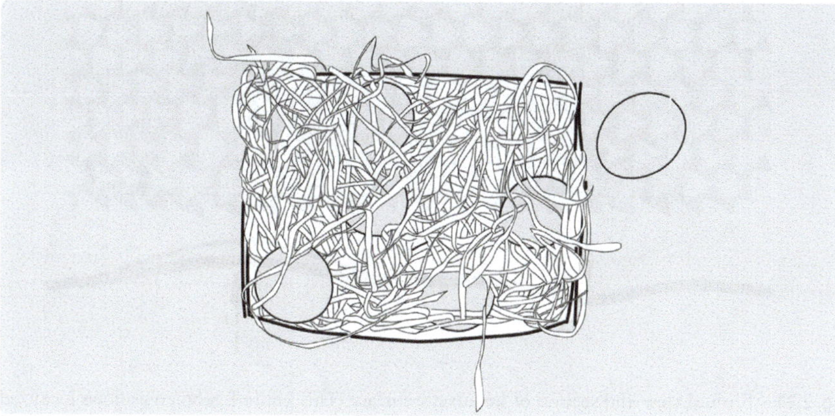

Fig. 2.19 Shredded paper packaging of a box with eggs. The crumpled and undulated paper surfaces are used as a dampening

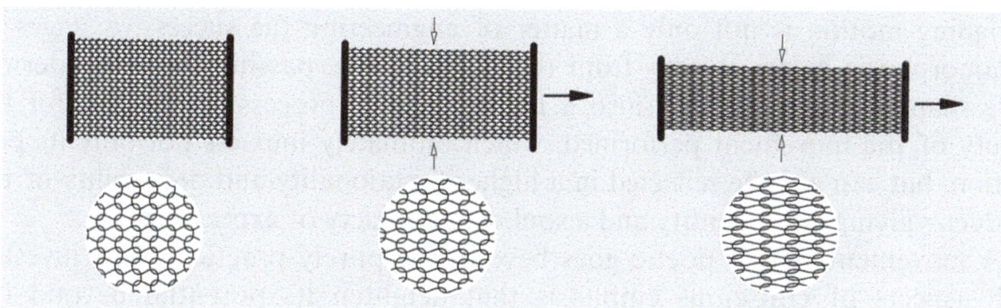

Fig. 2.20 Extension and thinning of a knitted surface with an enlargement showing the deformation of the pattern

- Change in stiffness. Structural stiffening of the structure can be modulated and controlled in detail by locally controlling the boundaries of the volume.

Expansion and *shrinking* geometries are capable of multiplying and dividing the volume they occupy by enabling change in one or more of their dimensions, in some cases also independently from change in the other dimensions. If these systems have a fix surface boundary, it needs to be very flexible to accommodate the change in volume (Fig. 2.21).

The main advantages for which this type of motion geometry is used are:

- Change in physical proprieties. Change in volume means change in the physical dimensions of a system and hence being able to modulate factors such as inertia, mass and shape.
- Sealing. Closing and opening of passages or views.

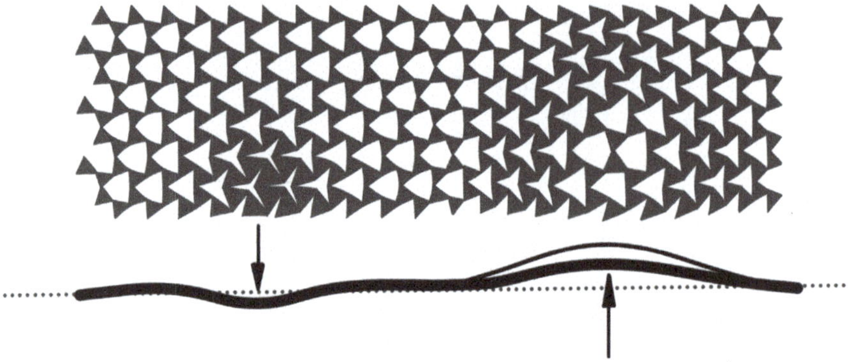

Fig. 2.21 Frontal view and section of an auxetic surface. This kind of geometry allows localized shrinking and expansion, accommodating to any underlying shape

2.4 Identity

Designing motion is not only a matter of engineering the successive stages of metamorphosis in an object—from the initial shape, passing through intermediate shapes, and back. It is also a highly creative process responsible for the beauty of the movement performed, which ultimately impacts not only its perception, but can also be reflected in a higher functionality and desirability of the product—giving it an identity and a soul with a variety of expressions.

A movement that is poetic goes beyond the purely practical. It is invested with aspects of conscious emphasis that heighten its potential beyond the purely technical. This apparent addition does not necessarily imply higher costs or lower efficiency, but what it can achieve if implemented successfully is an increase in usefulness and inherent value—becoming part of our cultural identity—and therefore also increasing its long-term sustainability (Schumacher et al. 2010).

Beauty does not come automatically as a by-product of its practical optimization but must be consciously embedded. A growing use is made of motion as a powerful and creative way of conferring anthropomorphic qualities to inanimate systems as a way to attract and keep the users' attention. In that sense, designers more than any other professionals in the field of the engineering sciences are trained to highlight the aesthetic potential of motion in objects, as is well represented in the smoothness of the movements and sounds connected to the design of modern vehicles in general. These are far from being casual and are instead the result of a very careful design and programming.

Without venturing into all the relevant multi-sensory aspects involved in the perception of motion, some factors impacting the aesthetic quality of the motion are briefly discussed. These concepts and principles are borrowed from the field of graphical animation, where a few of these key concepts have been identified and used for more than a century (Thomas and Johnston 1997).

Staging relates to the way the kinetic concept is presented. It is a way for the designer to direct what is being communicated to the observer by stressing one or more aspects of a movement while putting other actions in the background. It is a way to catch the observer's attention and help him focus on the movement so that part of the kinetic effect is not lost.

Emphasizing motion is an effective way of communicating to the user the physical proprieties of the bodies involved in the movement. There are many possible ways of emphasizing a movement. Most techniques consist in slightly exaggerating certain effects to give additional physical information. One such way is, for instance, the design of vibration and sound, which are among the major qualities in the attributes that customers look for in a sports car. These are part of the communication of power and reliability to the driver (Gupta et al. 2016).

Anticipation introduces a small action in the opposite direction of the main movement in the instants before it starts. It gives the observer a clue of what is going to happen next and in this way enhances the movement.

Supporting actions. If the "primary movement" refers to the main action performed (a bird flying), the "supporting actions" are all other movements that happen as an effect of the primary action (bird's feathers moving). Although the autoreactive mechanical design might be focused on the main action, secondary details and characters can be added to the system to highlight the overall qualities of the movement.

2.5 Conclusions

This chapter has reviewed a series of aspects that are central to an efficient morphological kinetic design. The parameters have been broken down into four main aspects that define the features as well as the effect of the kinetic action. These aspects are:

- The morphospace or the special trajectories that a body's shape follows through space and time. It is defined as the virtual frame of reference and the conventions used to describe the transition between morphologies;
- The morphology, or the geometrical categories of features the moving parts are made of;
- The morphosis, which refers to the process of shape change in time and space. Specific geometries which accommodate and embed an intrinsic potential for change are discussed with their strengths;
- The identity, which is an additional quality that can be conferred through designing a specific movement. It is responsible not only for the beauty of the movement performed but can also give access to a higher functionality.

As a brief inventory of conceptual tools, these single aspects need to be combined and skilfully matched with the aspects described in the other chapters in order to achieve smart and efficient autoreactive designs.

References

Arthur W (1997) The origin of animal body plans: a study in evolutionary developmental biology. Cambridge University Press, Cambridge, UK. ▶ https://doi.org/10.1017/CBO9781139174596

Attia S (2016) Evaluation of adaptive facades: the case study of Al Bahr Towers in the UAE. In: Q science proceedings, Qatar green building conference 2016 the action, vol 2016, no 8. Hamad bin Khalifa University Press (HBKU Press). ▶ https://doi.org/10.5339/qproc.2016.qgbc.8

Biewener AA (2007) Animal locomotion. Oxford University Press, Oxford, UK. ISBN 978-0-19-850022-3

Bush BMH, Clarac F (1985) Introduction. In: Bush BMH, Clarac F (eds) Coordination of motor behaviour. Cambridge University Press, Cambridge, pp 1–7. ISBN 978-0521264259

Cruse H, Graham D (1985) Models for the analysis of walking in arthropods. In: Bush BMH, Clarac F (eds) Coordination of motor behaviour. Cambridge University Press, Cambridge UK. ISBN 978-0521264259

Genchi GG, Marino A, Tapeinos C, Ciofani G (2017). Smart materials meet multifunctional biomedical devices: current and prospective implications for nanomedicine. Front Bioeng Biotechnol **5**(80). ▶ https://doi.org/10.3389/fbioe.2017.00080

Gipe P (2009) Wind energy basics: a guide to home and community-scale wind-energy systems, 2nd edn. Chelsea Green Publishing. ISBN 9781603582278

Gruber P (2011) Biomimetics in architecture, architecture of life and buildings. Springer, Wien. ISBN 978-3-7091-0331-9

Gupta A, Sharma S, Narayan S (2016) Vehicle noise and vibration. In: Gupta A, Sharma S, Narayan S (eds) Combustion engines: an introduction to their design, performance, and selection. Scrivener Publishing LLC, pp 103–119. ▶ https://doi.org/10.1002/9781119284543

Hardesty L (2015) Centimeter-long origami robot. Controlled by magnetic fields, tiny robot climbs inclines, swims, and carries loads twice its weight. MIT News, MIT News Office, 12 June 2015. Retrieved 12 Apr 2020 from ▶ http://news.mit.edu/2015/centimeter-long-origami-robot-0612

Hudson PE, Corr SA, Wilson AM (2012) High speed galloping in the cheetah (Acinonyx jubatus) and the racing greyhound (Canis familiaris): spatio-temporal and kinetic characteristics. J Exp Biol 215:2425–2434. ▶ https://doi.org/10.1242/jeb.066720

ICD (2013) HygroSkin: meteorosensitive pavilion, ICD research buildings/prototypes. Institute for Computational Design and Construction. Retrieved 21 Apr 2020 from ▶ https://www.icd.uni-stuttgart.de/projects/hygroskin-meteorosensitive-pavilion/

Ji A, Lei Y, Wang J, Ni Y, Dong B, Dai Z (2014) Variation in spatial and temporal kinematics of level, vertical and inverted locomotion on a stinkbug Erthesina fullo. Chin Sci Bull 59(26):3333–3340. ▶ https://doi.org/10.1007/s11434-014-0310-x

Lee D-Y, Kim J-S, Koh J-S, Cho K-J (2013) The deformable wheel robot using magic-ball origami structure. In: Proceedings of the ASME 2013 international design engineering technical conferences and computers and information in engineering conference IDETC/CIE. American Society of Mechanical Engineers, New York

Leman M, Moelants D, Varewyck M, Styns F, van Noorden L (2013) Activating and relaxing music entrains the speed of beat synchronized walking. PLoS ONE 8(7). ▶ https://doi.org/10.1371/journal.pone.0067932

Maiorana VC, Van Valen LM (2020) Bilateria: an organ-level of organisation. Animal. Encyclopaedia Britannica, Inc. Retrieved 10 Apr 2020 from ▶ https://www.britannica.com/animal/animal/Animal-diversity#ref405227

Maletic V (1987) Body, space, expression: the development of Rudolf Laban's movement and dance concepts. de Gruyter W (ed). ISBN 9783110107807

Marey EJ (1884) Geometric chronophotograph of a man in a black suit. In: Jensenius AR (ed) Some video abstraction techniques for displaying body movement in analysis and performance, vol 46, issue 1. Journal of International Society for the Arts, Sciences and Technology, Leonardo, pp 53–60 (2013)

Ogele TA (2013) Anatomical terms, anatomical planes. OpenStax, Rice University. Retrieved 7 Apr 2020 from ▶ https://cnx.org/contents/Gko70fNo@1/Anatomical-Terms

Peraza Hernandez E, Hartl DJ, Lagoudas DC (2019) Active origami. Modeling, design, and applications. Springer, Cham, Switzerland. ISBN 978-3-319-91865-5

Persiani SGL (2018) Biomimetics of motion, nature-inspired parameters and schemes for kinetic design. Springer, Cham, Switzerland. ► https://doi.org/10.1007/978-3-319-93079-4

Persiani SGL, Molter PL, Aresta C, Klein T (2016) Mapping of environmental interaction and adaptive materials for the autoreactive potential of building skins. In: Proceedings of the 41st IAHS World Congress on Sustainability and Innovation for the Future, Albufeira, Portugal

Pfeifer R (2000) On the role of morphology and materials in adaptive behavior. In: Meyer J-A et al (eds) From animals to animats 6, proceedings of the sixth international conference on simulation of adaptive behavior. The MIT Press, Cambridge, UK. ISBN 978-0262632003

Schmidt-Nielsen K (2004) Scaling, why is animal size so important? 1st edn., 1984. Cambridge University Press, Cambridge, UK. ISBN 0-521-31987-0

Schumacher M, Schaeffer O, Vogt MM (2010) Move. Architecture in motion—Dynamic components and elements. Birkhaüser, Basel. ISBN 978-3764399863

Steadman P, Mitchell L (2010) Architectural morphospace: mapping worlds of possible built forms. Environ Plan 37:197–220. ► https://doi.org/10.1068/b35102t

Thomas F, Johnston O (1997) Illusion of life: disney animation, disney editions, pp 47–69. ISBN 978-0786860708

Valentine JW (2004) On the origin of phyla. The University of Chicago Press, Chicago and London. ISBN 0-226-84548-6

Várkonyi PL, Domokos G (2007) Imperfect symmetry: a new approach to structural optima via group representation theory. Int J Solids Struct 44(14–15):4723–4741. ► https://doi.org/10.1016/j.ijsolstr.2006.11.047

Wolpert L, Tickle C, Lawrence P, Meyerowitz E, Robertson E, Smith J, Jessell T (eds) (2011) Principles of development, 4th edn. Oxford University Press. ISBN 978-0199549078

Mechanics of Autoreaction

Contents

S. G. L. Persiani, *Design of Autoreaction*,
Design Science and Innovation, https://doi.org/10.1007/978-981-15-6178-8_3

Autoreactivity fundamentally relies on an optimization of design features in order to achieve latent energy kinetics. This involves that the available energy is transmitted and multiplied in a convenient way so as to achieve a reaction even with the smallest energetic input. It is the field of mechanics that traditionally describes the transmission and optimization of the energy transformation within a specific movement.

Mechanics is the area of science concerned with the behaviour of physical bodies when subjected to forces or displacements. The study of mechanical principles and interactions is essential to understand how the physical assembly of a system's parts, in terms of geometry and physical proprieties, impacts the intrinsic potential of the system to convey energy and achieve motion along pre-defined kinetic patterns.

A perpetual motion machine is a hypothetical machine that can do work indefinitely without an energy source. It is the holy grail of mechanics, and if it were to be achieved, it would produce perpetual motion once activated through an energy input. The most common design for a perpetual motion device involves a wheel of some sort, often overbalanced so that it continually rotates using gravity as the driving force. However, there is a scientific consensus that perpetual motion in an isolated system violates either the first law of thermodynamics, the second law of thermodynamics or both. These state that energy cannot be created nor destroyed in an isolated system and that the entropy in the system always increases.

Although perpetual motion machines are not possible, a number of compatible principles can be used in combination to optimize the use of energy involved in the motion.

As previously discussed, autoreactivity in an artefact can be achieved on three different levels:

1. Autoreactivity is achieved on a system level. Non-dynamic parts form together a mechanism in which the interaction between the parts achieves the movement. If one or more parts are missing or do not perform correctly, the system as a whole might fall short of its autoreactive capacity.
2. Autoreactivity is embedded inside one or more materials composing the parts of the system. The parts of the system that do not have themselves any autoreactive potential are generally aimed at supporting and emphasizing the autoreactive capacity of the material. In this case, the parts realized with autoreactive materials keep their reactive capacities regardless of the functioning or presence of the non-reactive parts.
3. Autoreactivity is a combination of the two previous cases. It is a mechanism combining non-dynamic mechanic parts and parts made of materials with embedded autoreactivity. If either the material or the systems' parts do not perform correctly, the system as a whole risks to fall short of its autoreactive capacity.

This chapter focuses on the first case, autoreactivity on a system level. It explores the kinetic potential of autoreactive structures in mechanical terms and how the body parts can be assembled to perform specific types of motion. The aim is however not to represent a traditional review of kinetic and mechanic principles, but to facilitate the application of the examined principles in a design context and to draw parallels with the aspects discussed in the other chapters (in particular ▶ Chap. 2 Geometry and ▶ Chap. 5 Materials).

To achieve that, the mechanical parameters are broken down to simple intuitive principles to simplify the design of high-potential kinetic systems already in the early stages of the conceptual design of the autoreactive systems. This chapter adopts an interdisciplinary approach mixing elements of basic knowledge, theories and anatomy from the fields of mechanics, biomechanics and systems thinking. Transmission, amplification and control are the main topics discussed, which need to find an integrated solution when designing autoreaction.

Biomechanics is a multidisciplinary field that studies how the anatomy of biological organisms affects their motion in terms of forces and the mechanics. Living organisms efficiently combine their anatomical features, their energetic resources and in some cases their cognitive abilities to achieve precise, efficient and highly skilful movements. Therefore, to understand the complexity of these systems, biomechanics brings together mechanics and design optimization through natural evolution, bridging cardiovascular and orthopaedic biomechanics, sport biomechanics and many more.

Systems theory is an interdisciplinary field that studies the concepts and principles that drive the conglomeration of interrelated and interdependent natural or man-made parts to form an organized structure with a defined purpose. Every system is characterized by its predictable patterns of behaviour, its spatial and temporal boundaries and how these factors are influenced by the surrounding environment. In the context of autoreactive systems, it is the understanding of the underlying principles of self-organization, self-adaptation and emergent behaviour that is of interest in the context of control mechanisms in complex systems.

3.1 Principles of Biomechanics

Identifying the causes of motion may be the most useful kind of mechanical information for determining what potential changes could be used to improve autoreactive design.

This section describes a few basic core principles of biomechanics reported by Knudson (2007) that are mainly used in kinesiology, but that can also be applied within a biological and mechanical context. The aim is to suggest a simplified but informed approach to the technical assembly of autoreactive parts, fostering the systematic use of basic mechanical principles to improve the intuitive kinetic optimization of autoreactive systems.

3.1.1 Force-Motion Principle

The force-motion principle is based on the law of inertia, which states that all objects have the inherent property to resist a change in their state of motion. The principle states:

> **Force-Motion Principle**
> Whether in a resting or moving state, a system will tend to remain in that condition unless an unbalanced set of forces creates a movement or modifies it.

For example, suppose a person is cycling on a straight bicycle lane and spots a pothole on the lane a few metres ahead. To avoid hitting it, the cyclist must change his linear state of motion by using different muscles to operate the bike and create a sideward movement (a change in direction and speed).

This principle applies to autoreactive systems in the measure that to bring the system from its state of rest—within a framework of balanced forces—to achieve a specific motion, a new set of forces that disturbs the balanced state must be introduced. It is important to notice the specific sequence of events: first unbalanced forces act, and then changes in motion can occur.

3.1.2 Force-Time Principle

The force-time principle is concerned with the temporal strategy of force application in movements. The principle is derived from the mathematical formulation of the impulse-momentum relationship.

Momentum and impulse describe relatively similar concepts—that use the same units as mass, time and distance—which are however fundamentally different and are calculated in two completely different ways. Momentum ($\boldsymbol{p}$) is a vector that describes the quantity of motion of an object in terms of its mass ($\boldsymbol{m}$) and velocity ($\boldsymbol{v}$), expressed in (kg · m/s):

$$\boldsymbol{p} = \boldsymbol{m} \cdot v$$

The momentum is an essential part in calculating the force, because force ($\boldsymbol{F}$) is equal to the rate of change of momentum, also expressed as mass multiplied by acceleration (**a**):

$$\boldsymbol{F} = (\boldsymbol{m} \cdot v)/t$$

$$\boldsymbol{F} = \boldsymbol{m} \cdot \boldsymbol{a}$$

The impulse ($\boldsymbol{J}$) takes into account both the force ($\boldsymbol{F}$) acting on the system and the duration of time (t) for which it acts, so to calculate the impulse the force magnitude must be known. In other words, it represents the change of momentum over a certain period of time and is expressed in (N · s).

$$\boldsymbol{J} = \boldsymbol{F} \cdot t$$

The force-time principle states that:

Force-Time Principle

A force applied over a longer period of time (large impulse) achieves a greater speed (change in momentum) than if the same force was used over a shorter time interval.

When a golfer swings a golf club, the club gains momentum and hence a specific amount of force. When the golf club hits the golf ball with a very short interval of collision, the ball with mass (*m*) undergoes a large acceleration (***a***) as the energy is transmitted from the club to the ball (although some energy is dissipated under the form of sound and heat). Suppose now that the golfer repeats the same action, but this time takes a larger swing than the first time. As the duration under which the force is applied on the club is longer, the club will be imparted with a larger impulse than the first time, resulting (in theory) in a longer shot (◘ Fig. 3.1.).

In the same way, a player catching a ball will be able to slow down the ball more (absorb more kinetic energy) if he intercepts the ball early, by applying force in the opposite direction of the motion over a longer period of time (Knudson 2007).

Among the limits of the application of the force-time principle are (1) accuracy and (2) material characteristics. In the first case, when high accuracy is needed, extra speed becomes of lower importance. In the golf example, the player will make a much smaller but accurate movement to putt the ball into the hole, in comparison with the big swing in the first shot. In the second case, some material characteristics may have specific operation timeframes that need to be respected for optimal movement. For example, when preparing for a jump, humans rapidly bend the legs before the propulsion. Isometrically contracting muscle fibres that are suddenly allowed to shorten function exactly like a rubber band. However, if the force is not rapidly released, the muscle recovers adapting to its new length unloading the spring effect (Alexander 2000). Therefore, in this case, if the bending movement is lengthened beyond the operation timeframe of the muscle, the jump will not be as efficient.

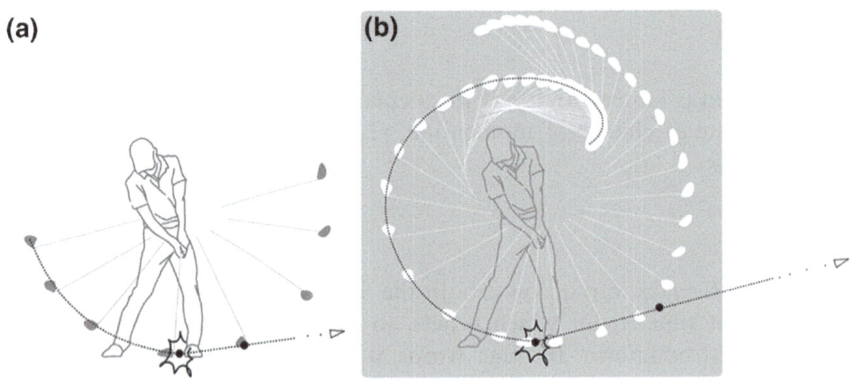

◘ **Fig. 3.1** Scheme of the movement involved in a low-energy (**a**) vs a high-energy (**b**) swing in a golfer. In the second case, the duration as well as the range of motion over which the movement is applied is much longer than in the first case, resulting in a higher energy and longer shot

3.1.3 Range of Motion Principle

The range of motion defines the amplitude and the complexity of the motion that can be achieved: which can be described defining the degrees of freedom and the width of the movement. As moving through a succession of body configurations (long or short) impacts the duration of motion, this principle is strongly connected to the force-time principle.

Range of Motion Principle
A small range of motion is effective to achieve high accuracy movements and goes together with low speed and low forces. A large range of motion favours high speed and high-force production or absorption but is not very accurate.

Living systems are able to manipulate their body structures so as to adaptively change their range of motion. For instance, many vertebrate animals can modify the number of joints and adjust the type and the amount of the joint rotations to tailor the range of motion. To limit the range of motion, parts of the body structure are stiffened while the moving parts can be operated with higher predictability: when throwing a dart, the person stabilizes the whole body focusing only on the movement of the elbow and wrist. To maximize the range of motion, the greatest moment the whole body can perform is used to impart lager force on to one specific segment or object (as also seen previously in ◘ Fig. 3.1.). This is the case of a javelin thrower who will use the whole body in movement and a long running approach to maximize the speed of javelin release (Knudson 2007).

3.1.4 Principle of Inertia

As previously mentioned, inertia is the property of all bodies to resist changes in their state of motion, whether at rest or moving the object or system will tend to continue whatever it is doing.

There are two numerical measures of the inertia of a body. In linear motion, the mass (m) of the body governs its resistance to the action of a force. In rotational motion, the moment of inertia (I) measures its resistance to the action of a torque around an axis. It is measured in (kg/m^2), as the sum of the products obtained by multiplying the mass (m) by the square of its distance (r) from the axis.

$$I = 2mr^2$$

Although the propriety is fundamentally a passive feature that does not enable the body to do anything except oppose any agent that comes to disturb its state, inertia can be used to an advantage when modifying motion or transferring energy from one body segment to another.

Principle of Inertia
Inertia can be modified by changing the quantity of mass in a system or by changing its distribution, resulting in a change of motion patterns.

3

Reducing mass to achieve linear acceleration. In linear motion, changing the quantity of mass involved in the motion changes its inertia allowing quicker changes in motion, like sharper turns, faster acceleration, etc. Birds with a smaller wingspan can make fast and dramatic manoeuvres, while large wingspans produce larger forces but are less manoeuvrable (Norberg 1981). In sports, this strategy is used in racing gear: while heavier shoes providing more stability are used in training, very light racing shoes make the athlete's motions lighter and faster (Knudson 2007).

Change in the distribution of mass in spinning motion. The formula for the moment of inertia shows that the body's resistance to rotation depends more on distribution of mass (r^2) than on the quantity of mass (m). While the mass of a body is constant, the same body has as many different moments of inertia as the number of axes the body can be rotated around. The pattern of a spinning movement can therefore be manipulated by modifying the body's orientation to the rotation axis or by changing the distribution of the body segments.

Decreasing inertia achieves rotational acceleration. Decreasing inertia by compacting the body mass close to the axis of rotation creates an acceleration in the spinning movement. This can be seen in athletes that pulling their extremities towards the body core to initiate a rotation (◘ Fig. 3.2, steps 1–5).

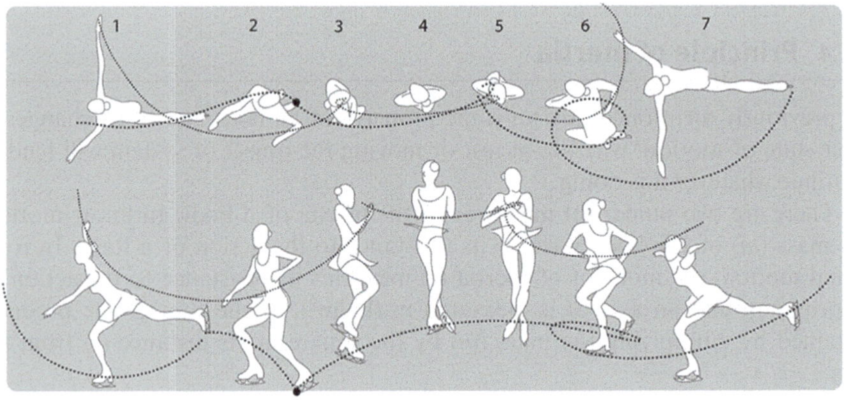

◘ **Fig. 3.2** Scheme of a figure skater performing a salchow jump. The movement of the right arm and the left foot is highlighted during the sequence. The athlete uses her arms and legs to maximize the range of motion (1) to then concentrate the kinetic energy gathered and jump by pointing the left skate tip into the ice (2). While jumping, the arms and legs are suddenly compacted around the spinning axis to decrease inertia (2–3) and drive the spin (3–6). Arms and legs are then spread again to increase inertia and allow a stable landing (6–7)

Increasing inertia achieves more force and better stability. Applying a weight far away from the rotation axis increases inertia and improves the force output as well as the stability of the movement (◘ Fig. 3.2, steps 5–7). For example, tennis players improve stability and force output by adding lead tape at the top of the racket frame.

Transfer of mass to other body parts. In many ball sports, the forward motion of a good percentage of body mass can be transferred to a smaller body segment just prior to impact or release, modifying the motion of other bodies. This strategy is also typical in martial arts, where defensive moves take advantage of the inertia of the attacker to redirect the blows (Knudson 2007).

3.1.5 Principle of Balance

Mechanical equilibrium occurs when the forces acting on an object sum to zero, resulting in an object either being at rest or moving at a constant velocity.

Stability and mobility are inversely related. Highly stable postures strongly resist changes in position and movement. Unstable postures on the other hand facilitate mobility, allowing to create or modify movement with a very little energy input. The capacity to move and change therefore strongly depends on unstable proprieties within a body structure. To achieve skilful movement however, the right mix of stability and mobility is needed.

Balance refers to the capacity to control a specific body position, whether in static equilibrium conditions or during dynamic movement, in relation to a defined base of support. Good balance can be achieved in highly adverse situations and does not always require good mechanical posture: the principle of balance is based on the mechanical trade-off between stability and mobility.

Principle of Balance
Postures with smaller bases of support, with the centre of gravity not too close to the base of support, foster higher mobility over stability.

Control of the stability/mobility relationship is easily achieved by modifying (1) the base of support and (2) the position of the centre of gravity in relation to the base of support.

- The base of support is the contact area between the supporting segments of the body and the medium (air, water, ground). The larger the base of support, the greater the area to distribute the bodyweight, hence the greater the stability.
- The distribution of the body and its parts determines the position of the centre of gravity relative to the base of support. If the body leans so that the line of gravity falls outside the base of support, motion is facilitated in that direction under the action of a gravitational torque. A centre of gravity that is more distant to the base of support can be moved beyond the base of support easier than in postures with a low centre of gravity.

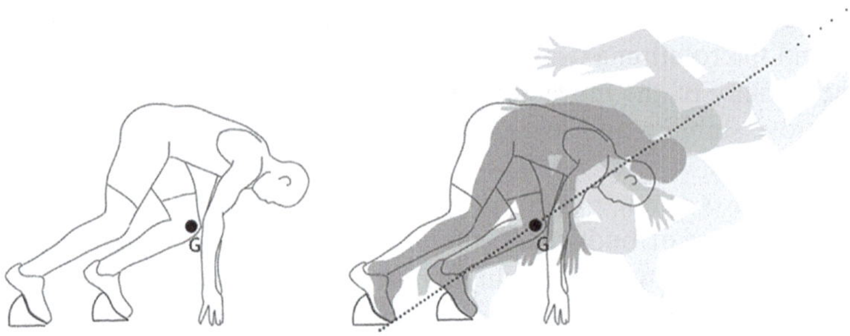

Fig. 3.3 Scheme of a racing athlete in the starting position. The athlete awaits the start holding the hips elevated to shift the centre of gravity (*G*) as high as possible and slightly forward of the base of support. When the gun is fired, the athlete lifts the hands from the track while applying force (*F*) against the blocks and through the body at an angle of approximately 45° (Harrison and Comyns 1997)

A typical example of a posture with less stability in favour of increased mobility is the starting position of athletes in a track race. As the direction of motion is known, the athletes take an elongated body posture leaning forward, with the centre of gravity near the edge of the base of support (Knudson 2007) (Fig. 3.3). This position allows, in comparison with a standing position, to rapidly take up a mechanically efficient running position and reach top speed as quickly and smoothly as possible.

3.1.6 Coordination Continuum Principle

The sequential coordination and timing of specific body part movements determine the kinematic outcome of the action. The coordination of parts can be simultaneous or sequential and is in most biomechanical movements a combination between both strategies (Knudson 2007).

In *simultaneous coordination*, actions are timed to be unleashed at the same time. This results in the summed effect of the action of many body parts, ultimately producing higher forces. For instance, a person lifting a heavy weight might simultaneously activate leg, arm and core muscles.

In *sequential coordination*, actions are timed to be unleashed one after another. This mainly results in an acceleration, ultimately producing high-speed movements. For instance, a person wishing to throw a weight vertically in the air will start the action with the legs, followed by trunk and arm motions.

The coordination continuum principle suggests that:

Coordination Continuum Principle

Movements aiming to generate high forces tend to utilize simultaneous segmental movements, while generation of lower force and high-speed movements involves a sequential movement coordination.

This principle typically develops intuitively with time in many biological systems. This is why inexperienced individuals can be observed to make attempts with the simultaneous action of only a few joints, while more experienced actors use more segments and greater sequential action to perform the same action, but with highly skilled movements.

3.1.7 Segmental Interaction Principle

The forces acting in a system of linked rigid bodies can be transferred through the links and joints so as to concentrate the energy of large muscle groups in specific body parts (Knudson 2007). For instance, a whip can be operated without the use of the adduction muscles by exploiting the intersegmental reaction forces. The operator accelerates the arm in the direction of the blow by extending it and then abruptly slowing it down as the peak speed is reached. This negative acceleration creates a backward force and a torque at the height of the wrist that positively accelerates the whip.

Segmental Interaction Principle
Forces acting between the segments of a body can transfer energy between segments.

3.1.8 Optimal Projection Principle

The optimal projection principle applies to movements involving projectiles and defines the range of angles of projection that result in the best performance. The optimal angle of projection in each action can be expressed as a specific compromise between vertical velocity and horizontal velocity. Vertical velocity is the factor that determines the duration of the flight, and the horizontal velocity the range of displacement.

Optimal Projection Principle
The angle of projection that, in geometrical terms, maximizes the horizontal displacement is 45°.

Angles under and above 45° create shorter displacements because of loss of flight time that cannot be compensated by the higher horizontal velocity or loss in horizontal velocity. When the aim of the displacement is accuracy rather than distance, the range of the optimal angle of projection is however reduced.

The 45° geometrical rule is however affected by other factors as the height of release in comparison with the height of the target, the biomechanical force of release and the resistance of the medium (usually air). Specifically, stronger release and air resistance make in most cases a lower angel of release more effective for a longer horizontal displacement (Knudson 2007).

3.1.9 Spin Principle

The spin principle also applies to movements involving projectiles and defines how the overall movement is affected when a spin or rotation is imparted to the object in order to obtain an advantageous trajectory or bounce.

Spin Principle
Spin is desirable upon thrown objects as it stabilizes the orientation of the object during flight ensuring aerodynamically efficient flight and can create fluid lift forces able to counter the existing forces involved and model the horizontal displacement.

Spin is created by imparting a force off-centred with respect to the centre of gravity of a body, creating a torque that brings the body to rotate. The linear speed of the projectile is however inversely proportional to the spin created. This means that more spin produced in the release comes at a cost of lower flight speed. Therefore, the spin needs to be balanced with the other forces involved to achieve optimal output.

The aerodynamic stabilizing effect of the spinning body is the result of applying the law of inertia: as any object in angular motion without external acting torques conserves its angular momentum, the projectile is able to hold a specific orientation as a result.

The fluid lift forces strongly depend on the type of spin and on the geometry of the object. In a ball, a topspin creates a downward lift force resulting in a steep diving flight, while a backspin creates an upward lift force that increases the distance of a drive (Knudson 2007).

Finally, as the body ends its flight and hits a surface that interrupts the movement, the spinning motion of the object can create complex bouncing and landing motions that are difficult to predict.

3.2 Actuators

The actuator is the component responsible for initiating motion in a system and is not to be confused with the agonist mechanism (discussed further down in the text), which is the mechanism conveying the first movement in the timeframe of reference. When we reach out to take an object, the actuator is our brain, which decides on the moment and the way to grab the object, and the agonist mechanism is the muscle that allows our arm to stretch out and the antagonist mechanism the muscle that allows our arm to be pulled back. To work, any type of actuator requires a control signal and a source of energy. When considering different types of actuators, most of those taken into account rely on active actuation.

In active actuation, the system purposely initiates a reaction in answer to, or to overcome, a disturbing factor. As a decision-making system, it generally requires a higher form of complexity enabling it to process multiple input–output combinations and decide on the outcome.

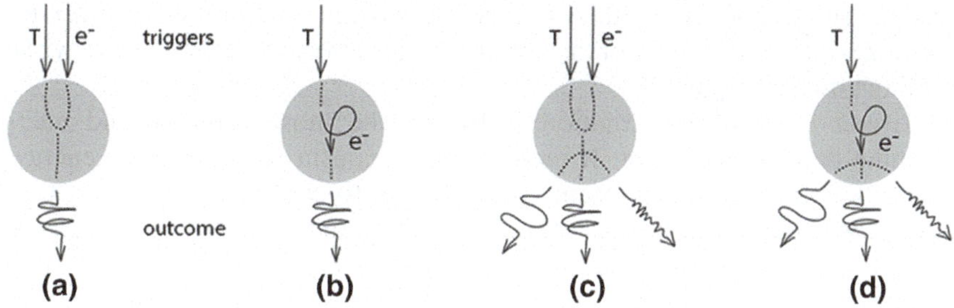

Fig. 3.4 Schematic representation of the operating contexts depending on trigger (*T*) and energy source (e−) in **a** P-E actuation (external energy source, one outcome per trigger); **b** P-I actuation (internal energy source, one outcome per trigger); **c** A-E actuation (external energy source, more outcomes per trigger); and **d** A-I actuation (internal energy source, more outcomes per trigger)

Passive actuation on the other hand is a much simpler system that requires little to no decision-making. The typical actuation-reaction pattern follows the same logical order as in a domino-effect: to one input (pushing the first domino), corresponds a predictable output (all dominos fall) with a predetermined succession of reactions (falling succession and timing depend on the specific layout pattern).

Further distinctions can however be made within the active and the passive actuation families, depending on the source of energy (external or internal) employed in the actuation action, which is highly relevant in an autoreactive context.

Passive *external* (P-E) actuation is the most elementary form of actuation, the source of energy is external, and actuation is carried out as a sequence of predetermined actions (Fig. 3.4a). These systems require (virtually) no other energy input than the energy received by the system from its environment. Action and reaction have a 1:1 relationship in real time as input/response/outcome. To trigger a new adjustment, the input must change (Fox and Kemp 2009). A typical example is the windmill, which operates under favourable atmospheric conditions and stops as the wind drops below the critical limit. This system is also used by many plants in the context of responsive reactions (nastic or tropic movements) (Persiani 2018).

Passive *internal* (P-I) actuation requires an internal source of energy and a sequence of predetermined actions following actuation (Fig. 3.4b). Most mechanical devices that incorporate a motor, whether electrical- or petrol-based, use this type of actuation. In natural systems, passive internal actuation is found in plants, simple mono- and multi-cellular systems, and the visceral functions of higher animals.

Active *external* (A-E) actuation involves that the energy for the actuation is provided by the surrounding environment, but that the system itself is in control of the output (Fig. 3.4c). This requires a higher internal complexity with a processing system able to decide on a varying output. In a man-made context, a typical example is the paragliding aircraft that relies on the reaction of the air against its lifting surfaces to achieve free flight but needs to be operated by a pilot who can take instantaneous decisions to modify the structure of the aircraft to ensure the success of the flight. A parallel example in a natural context are birds exploiting winds and rising hot air currents to move over wide territories without providing powered flight.

Active *internal* (A-I) actuation is by far the most complex and energy-consuming type of actuation. It requires the organism to provide the energy to power action and a complex brain sensor system to control the outcome as well (▣ Fig. 3.4d). Examples in the technical world include self-driving vehicles and other automated systems. In nature, these systems are found in animals and do not appear to exist in plants. As autoreactive systems rely on external energy sources, this category is not further explored.

In the context of autoreactive systems, P-E and A-E actuations are of interest as these rely on external energy sources. P-I and A-I actuations are not further explored.

P-E actuation is the most common and straightforward form of actuation for autoreactive systems, whether actuated on a system level through a mechanism or on a material level through an autoreactive material. The system requires low complexity and an embedded sequence of reactions to perform the set functions.

A-E actuation is, in principle, not to be excluded in an autoreactive context, but it requires a complex controlling mechanism that has not been identified in complete autoreactive systems so far. It is however possible that future technological ephemeralization enables the development of A-E autoreactive systems as well.

3.3 Agonist Mechanisms

Because most mechanisms cannot actively produce both movement and countermovement, force must be transmitted so as to return the mechanism to its initial state. Either the mechanism is shaped in such a way that the movement is able to reverse itself without the need to provide a counteraction, as in many circular or looped mechanisms, or an action mechanism must work in pair with a reversing mechanism. Autoreversing mechanisms based on circular transmission systems (as wheels, belts and gears) are quite common in man-made applications but are very rare in natural systems that instead rely on other techniques to reverse the position of their body parts.

The mechanism that brings the system out of its initial state of equilibrium is typically called the *agonist*; the mechanism responsible for bringing the system back to the initial position is the *antagonist* (Biewener 2007). Agonist mechanisms are coupled in direct succession to the actuator and are mainly responsible for reaction.

The mechanics involved in the agonist systems are first described by discussing the different types of transmission mechanisms, whether using a soft-bodied system or a rigid body. The types of mechanisms are then classified according to the type of movement they achieve.

3.3.1 Transmission Mechanisms

The choice and the combination of the physical and mechanical proprieties of the body parts involved in motion are central aspects to effectively transmit energy and actuate a mechanism (Nachtigall 2005), whether it is of biological or synthetic nature. Geometries and transmission techniques are combined in distinct patterns depending on whether the system relies on principles of (1) rigid or (2) deformable bodies. In fact, all real bodies are deformable and will adapt under the pressure of forces according to its elastic and plastic proprieties. However, in classical statics and mechanics, the body is considered as rigid and incapable of deformation when the deformations are small enough to be neglected.

Rigid bodies are mostly used for their structural potential—providing support and force amplification—or as protective elements—where permanent deformation is designed to absorb shocks. The kinetic outcome of the actuation of rigid body parts can be described as a series of translational and rotational movements (Shigley and Uicker 1995). This type of mechanics is largely used in man-made kinetic systems as it allows dynamic control and material resistance to wear. Rigid body mechanics are also broadly used in nature, both by plants and by animals. Vertebrate organisms, for instance, use their internal skeleton as a lever system operated by muscles to maximize the amount of force applied to the external environment.

Deformable, or soft, bodies use their flexible and elastic material proprieties to retain and store to some extent part of the forces acting on the body, to reversibly return (completely or partially) to the initial form or state once these forces are removed. The motion of deformable bodies strongly depends on the internal stresses and strains of the materials, their working limit and their mechanics of collapsing. As man-made elastic materials developed up to today tend to lose their capacity to reversibly deform over time, systems using deformable body mechanics are quite uncommon in traditional engineering contexts. In fact, the deformable proprieties of a material are generally considered as a weakness connected to structural collapse. Elastic materials are sometimes integrated within rigid systems to amplify the kinetic effect, but macro-systems relying on soft body mechanics are rare. In nature on the other hand, flexible mechanisms are extremely common.

In fact, in nature both mechanical logics are often integrated, being opposite but complementary, to achieve very adaptable solutions as well as multifunctional complexity. Rigid mechanical bodies are integrated with flexible features, and soft-bodied mechanics with stiff ones: the exchange between the mechanical stiffness and flexibility is in most cases key element in the movement (Persiani 2018).

3.3.1.1 Rigid Systems

In rigid systems, mechanical transmission is not achieved through the deformation of the body structure. On the other hand, rotation and translation of pre-shaped systems are used to convey motion without altering the geometric proprieties of the parts.

In nature, vertebrates use the segmented subdivision of their internal bearing structure to allow rigid transmission of movement in pre-defined directions. Synthetic man-made systems are on the other hand for a very large part made out of rigid parts. In both contexts, the specific design of the parts allows to closely control the movement patterns by embedding all constraints and possibilities within the system. The shape of the vertebrae as that of the mechanical parts is adapted to every specific system to optimize posture, resistance and movement (Nachtigall 2005).

Levers and Linkages

A *lever* is a movable rigid beam that is attached to a fixed point and pivots on a fulcrum. As the input force is applied at a greater distance from the fulcrum than the output force, the mechanism acts as a force amplifier. As power is the product of force and velocity, a force applied to a point farther from the pivot needs less force to produce the same power as a force applied to a point closer to the pivot (Shigley and Uicker 1995).

Three different kinds of lever systems can be distinguished, depending on the order of the effort, the load and the fulcrum (Persiani 2018; Sclater 2011).

- Fulcrum in the middle. The effort is at one end and the load at the opposite end, as in a pair of scissors or a playground seesaw (the seesaw's contact with the ground is the fulcrum, the upper person is the effort, and the lower person is the load) (Fig. 3.5).
- Load in the middle. Fulcrum is at one end and the effort at the opposite end, as in a nutcracker, a wheelbarrow or a simple bottle opener (the bottle cap is the fulcrum, the part of the cap opening the load and the hand the effort) (Fig. 3.5).
- Effort in the middle. Fulcrum is at one end and load at the opposite end: as in a pair of tweezers or in a fishing rod (wheel is the effort, rod resting against the fisherman's trunk is the fulcrum, and fish is the load) (Fig. 3.5).

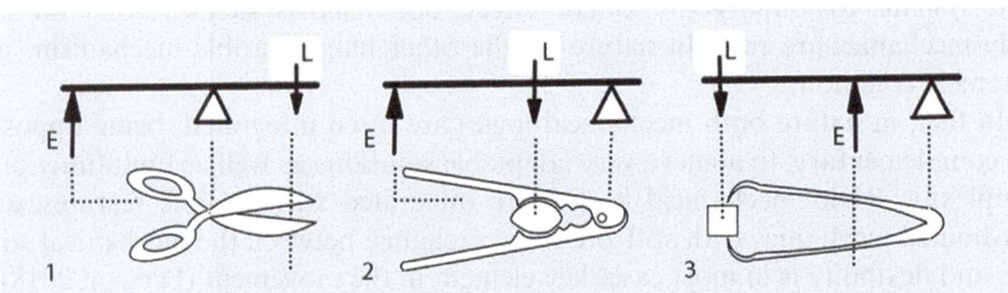

Fig. 3.5 Schematic representation of three types of lever systems with correspondent examples: (1) fulcrum in the middle (a pair of scissors); (2) load in the middle (a nutcracker); and (3) effort in the middle (a pair of tweezers)

Linkages are mechanical assemblies consisting of two or more levers connected to produce a desired motion, such as changing the direction of a force, driving the movement of more objects at differing speeds or performing straight lines and curves.

Rigid body structures in animals, as in the human skeleton, make a smart and adaptive use of their musculoskeletal system achieving to adapt the same structure to different purposes. By activating different sets of muscles, some parts of the system can be made more or less rigid, allowing the organisms to use their body parts as levers or as linkages depending on the need.

Inclined Planes and Screws

Inclined planes are used to produce mechanical advantage by applying a smaller force over a longer distance to lift a load vertically: the mechanical advantage equals the length of the slope to the height it spans. The longer the inclined plane to overcome a set height, the smaller the force needed. On the other hand, a shorter plane will require more force input but can achieve a faster result.

Inclined planes can be found in many different objects with a great variety of applications and uses (◘ Fig. 3.6). A few are further discussed:

Inclined surfaces are broadly used in the natural world as well as in man-made objects to produce mechanical advantage or thrust, as in the case of many different kinds of airfoils. Sharks and birds have asymmetrical wedge-shaped pectoral fins/wings producing lift that keeps the animal afloat and glide forward (Biewener 2007). The same shape is imitated in airplane wings and in hydrofoil boats.

A *wedge* is a triangle-shaped tool that is used to lift (such as a ramp) or to separate two objects (such as an axe).

A screw is made out of a narrow and inclined plane wrapped around an axle. In the technical context, the screws are used to achieve mechanical advantage by digging into a material producing resistance by wedging into it when turned, and also providing a fastening effect by preventing a linear pull-out. Screws can also be used to move large loads with small effort, as when a single person uses a car jack to lift a car.

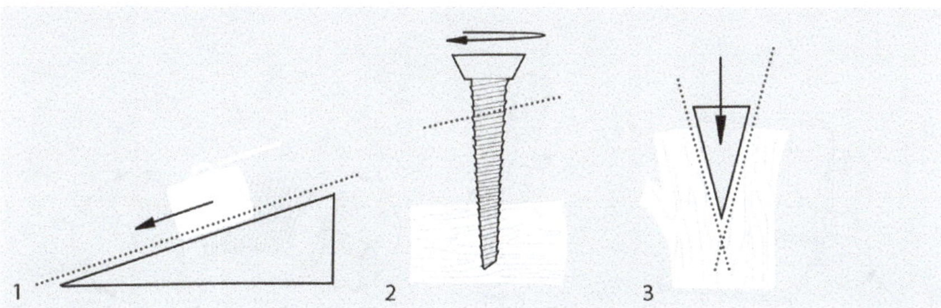

◘ **Fig. 3.6** Schematic representation of three different inclined planes: (1) inclined plane; (2) a screw; and (3) a wedge

Joints

Joints (in biomechanics) and linkages (in mechanics) are the connecting parts between rigid elements. The design of the joint not only allows to control movement, by designing with which timing and in which direction forces should be transmitted—by fixing degrees of freedom—it also allows mechanical advantage. Rigid structures without joints are generally used for protection (as shells) and cannot really be considered to be part of the kinetic actuation.

The design of a joint defines its mobility and the amount of force that can be transmitted across it. Simple joints with limited mobility can often transmit large forces than joints with high mobility.

Pivot joints transmit a rotating motion and can transmit large forces. Either a pivot rotates within a ring, or a ring moves around a pivot.

Sellar joints transmit swing movements with rotation of the body, which are typical of flexion–extension and abduction–adduction movements. The geometry of the joint is convex in one direction and concave in the direction at right angles to the first.

Hinge joints connect two rigid parts allowing a limited angle of rotation around a fixed axis (◘ Fig. 3.7).

Ball joints allow free movement in two planes at the same time, including rotation. Combining two such joints with control arms enables motion in all three planes (◘ Fig. 3.7). Ball joints can be spheroidal or ellipsoidal. *Spheroidal ball joints* ("ball-and-socket") allow three degrees of freedom. The spherical surface of the male element moves within a depression on a female element. *Ellipsoidal ball joints* have only two movements and allow opposition only to a small degree. They are made of two ovoid surfaces, which are both either convex or concave.

Flexure joints are as common in nature as they are rare in technology: they rely on the bending of the material or the structure in one specific point. In most cases, the material of which the flexure joint is built needs to be able to repeatedly bend without showing wear and deterioration (◘ Fig. 3.7). Applications of

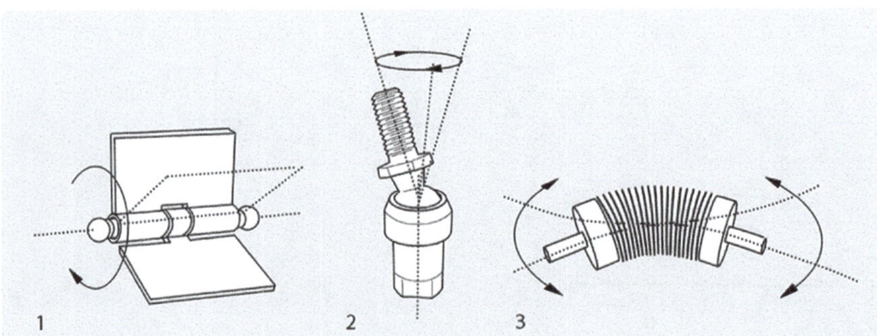

◘ **Fig. 3.7** Schematic representation of three common types of joints: (1) a hinge; (2) a ball joint; and (3) a flexure joint

flexure joints in mechanical contexts include variable stiffness actuation (VSA) in robotics, where the stiffness of the movements in the machines is regulated by introducing elastic elements into the joints (Petit et al. 2010).

Gears

Gears are wheels or cylinders with evenly spaced and formed teeth around their perimeter, which can be used to transmit motion and to create mechanical advantage. When two or more gears are connected to transmit motion, the mechanism is called a gear train. Gear trains are versatile and simple mechanical systems that are broadly used in mechanics and can among other actions perform changes in rotational speed and direction, converting rotational to linear motion Gear trains are versatile and simple mechanical systems that are broadly used in mechanics and can, among other actions, perform changes in rotational speed and direction, converting rotational to linear motion or reversely, linear to rotational (Sclater 2011).

In nature, gears are very rare, and the first functional biologically gearing system was discovered in the hopping insect *Issus coleoptratus* which helps it to coordinate its hindlegs as it jumps (Burrows and Sutton 2013).

Chain Systems

Chain systems are made of rigid elements or bodies connected to each other by joints forming a closed chain or a series of closed chains. When one link is fixed, the possible movements of the chain relative to (and the movement of each link relatively to one another) depend on the number of links and the type of joints. The mechanism can be designed to convert uniform rotary to non-uniform rotary or to oscillatory motion by changing the length of the links (Encyclopædia Britannica 2012).

Roller chains are commonly used for transmitting mechanical power (as the chain in a bicycle). They consist of a series of short links held together by hinges that allow the links to rotate relatively to one another only bi-dimensionally.

Kinematic chains assemble rigid parts with different and potentially complex joints, allowing very complicated three-dimensional movements. Terms as "open" or "closed" kinetic chains address whether the distal segments in the chain are restrained by external resistance or not.

In biomechanical terms, the linked segments of the human body can be considered as a kinematic chain. Unconstrained movements as arm or leg extension are classified as open chains, while constrained movements as leg squats are classified as closed chains (Knudson 2007).

Pulleys and Belt Systems

Pulleys are wheels or cylinders (gears without teeth) that are used in combination with a belt, chain, rope or cable to change the direction of a force and the speed of a rotating movement and achieve mechanical advantage. A combination of pulleys can produce considerable mechanical advantage. As with gears, the velocity transmitted by a pulley is inversely proportional to its diameter.

Pulley systems are simple mechanisms that are very common in man-made contexts but have not been found in nature.

3.3.1.2 Soft-Bodied Systems

Soft-bodied (or hydrostatic) systems in nature achieve mechanical amplification without using rigid skeletal elements, but by locally deforming their elastic body through muscle contraction. Variations of systems using these same principles are widely used in plant cells and in soft-bodied organisms but can also be found in secondary body parts of vertebrate organisms (Kier 2012).

3

Most of these systems are built with an elastic envelope containing diverse arrays of muscle bundles. The internal parts contained by the envelope are subdivided into one or more compartments with an incompressible fluid. The internal fluid is used; because of its very specific material proprieties: high bulk modulus and small shear modulus, the material is hard to compress and easy to deform in five out of the six principal directions,[1] allowing deformation at a constant volume. The incompressible proprieties are used as a vehicle for movement by allowing selective deformation and displacement of the fluid depending on the scaling and deformation of the elastic envelope. Thus, by geometrically deforming the body, a various array of movements and changes in stiffness can be reversibly achieved within defined ranges of stress. The power amplification effect is usually the combined outcome of a deformation of the compartment(s) and the elastic flexibility of the compartments' walls (Persiani 2018).

The geometrical composition and the number of compartments involved in the movement—whether the compartment is single, multiple and connected to each other, or multiple and juxtaposed—strongly affect the mechanics of the system. Although variations of compartment systems are found in most organisms, whether as body structure or for internal functions, single and multiple compartments have not been found to work in combination.

In a man-made context, systems relying on soft-bodied mechanical proprieties are quite uncommon, in comparison with the number of rigid body systems used. This is mainly due to the lack of reliable materials able to sustain repetitive and high-load mechanical stresses and strains without showing signs of wear or loss of the mechanical proprieties.

Single Compartment Systems

Single compartment systems often need to interact with the surrounding environment to achieve motion. A single pocket or body cavity of cylindrical shape is typically inflated, filling up with water or air from the exterior environment. Once the body cavity has been closed, the flexible body can be deformed to some extent by selectively tightening some muscles and relaxing others.

Deformation at constant volume means that since the volume (V) of the cylinder remains constant, and V being dependent on the cylinder's length [l] and radius [r],

---

1 The six directions of deformation of a body in a three-dimensional coordinate system can be described by defining the axis along which the deformation occurs (x, y, z) and the direction of deformation (+; −).

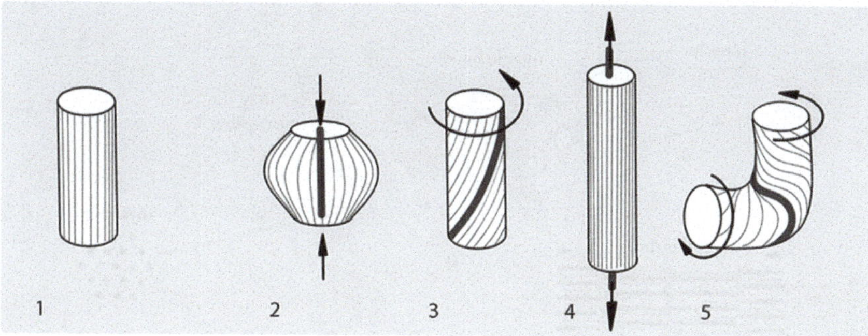

Fig. 3.8 Schematic representation of kinetic movements achieved with the deformation of a single compartment system with a constant volume: (1) undeformed compartment; (2) compression; (3) twisting; (4) extension; and (5) bending

$$V = \pi r^2 l$$

as the radius of the cylinder is changed upon compression of the muscles, the equation can be written with the radius differentiated with respect to the length (Kier 2012):

$$\frac{\delta r}{\delta l} = \frac{-r}{2l}$$

The effect of this type of deformation can, for instance, be seen in a water-filled balloon, which deforms when squeezed (Fig. 3.8).

Mostly however, these systems are used to achieve fast and powerful displacements. In this case, the filled compartment is put under increasing pressure through muscle compression until it eventually bursts, releasing the content under the form of a jet. The kinetic energy released by the jet projects the organism in the opposite direction to the jet with an explosive-like movement pattern.

This effect is the same that propels blown-up balloon through the air, as it is released without closing the opening.

Juxtaposed Compartment Systems

Juxtaposed compartments are a variation of the single compartment system, which however do not rely on the explosive release of the internal liquid.

In this case, many single compartments are positioned next to each other without being internally connected to each other. The kinetic energy in the deformation of one compartment is transmitted to its juxtaposed compartments through physical contact. If all the compartments react at the same time, the accumulation of the kinetic effect of each and every single compartment on a microscale results in a kinetic actuation of the system on a macro-scale.

This type of mechanics is observable in many natural structures on a microscopic scale. It is at the base of the movement in muscle fibres in animals and can

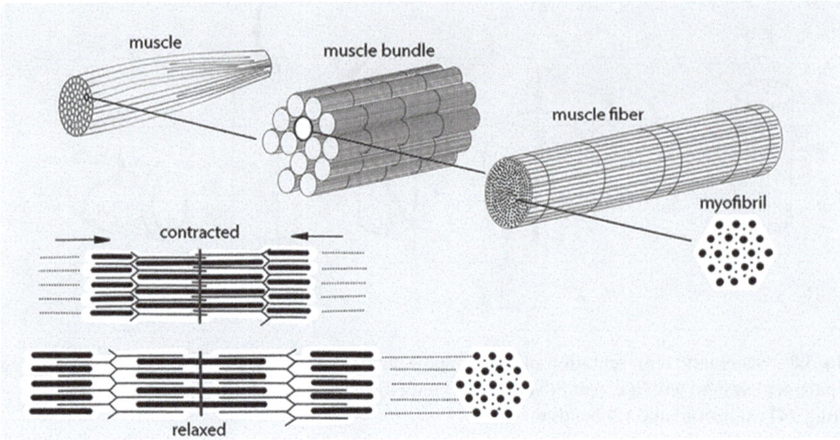

Fig. 3.9 Schematic representation of the structure of a muscle. The myofibril is a basic rod-like unit of a muscle cell. Muscles contract by sliding the thick and thin filaments (bottom left) along each other

be found in many hygroscopic mechanisms in plants as the vegetal fibres absorb and release water allowing the plants to change shape in different water conditions (Persiani 2018) (Fig. 3.9).

Multiple Compartment Systems

Multiple compartment systems are juxtaposed compartments that are internally connected, allowing a much more complex array of movements than in the single compartment system. The geometries of movement achieved depend on the spatial organization of the compartments and on the design of the bonds between the compartments. In animal systems, the pressure in each compartment is usually individually controlled, so the displacement of the fluid from one compartment to the next can be closely manipulated. These systems are therefore commonly used by soft-bodied organisms as hydrostatic skeletal support, to create movement and exert force.

The cylindrical geometry used by most of these systems works towards the mechanical advantage of the system. The radial control of the cylinder allows control and elongation of the cylinder in the longitudinal direction. The circumferential stress (C) is twice the longitudinal stress (L) in a pressurized cylindrical vessel, with pressure (p), radius (r) and thickness (t):

$$C = \frac{2pr}{2t} = 2L$$

Typical movement patterns used by multiple compartment systems are sequential wave-like activations of the single compartments producing metachronal or peristaltic propulsion waves (Fig. 3.10).

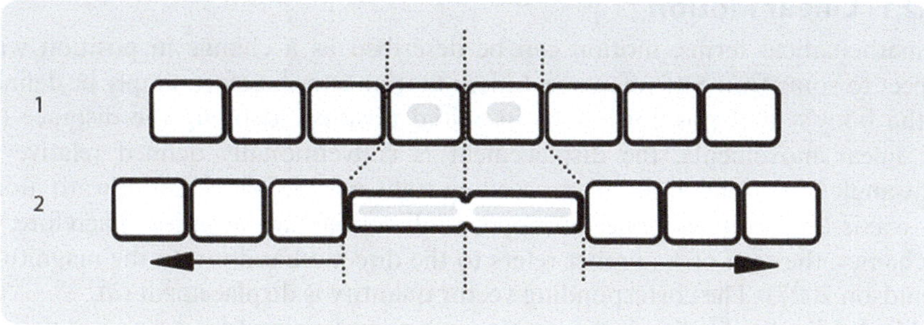

Fig. 3.10 Schematic representation of the kinetic movement due to the actuation of a multiple compartment system: (1) undeformed state of the succession of compartments and (2) effect of the squeezing of two compartments. This basic principle is used, for instance, by earthworms, but under the form of a propulsion wave

3.3.2 Types of Movement

The definition of the mechanical components in a motion control system is of primary importance in the overall design of a system, whether it is of an auto-reactive or a powered type of machine. This is because the mechanics involved ultimately determine the operation and maintenance requirements of the system, and therefore also any additional electronic circuitry in terms of motion controller and software (Sclater 2011).

The term *machinery* refers to an assembly of machines and mechanisms. These two terms are often mixed up, although they refer to quite different concepts. A machine transforms energy to do work, as in a combustion engine. The assembly of the piston, the rod and the crankshaft composing the slider-crank within the combustion engine that transmits the motion is a mechanism (Sclater 2011).

There are many possible ways of classifying mechanisms. One classical way is to classify mechanisms depending of their range of mobility.

Planar mechanisms move in a bi-dimensional space, and the movement of all particles within the mechanism can be described as moving in parallel to a single common plane. Planar mechanisms are simple to design as all movements can be represented in true size and shape from a single viewing direction or the same reference plane.

Spherical mechanisms typically have one point which remains stationary, while the other points' paths lie on concentric spherical surfaces. Spherical linkages are all revolute pairs, with the axes of all revolute pairs that intersect at a point, as in universal joints and bevel gears.

Spatial mechanisms must be described in a three-dimensional coordinate system or with non-Euclidean geometry as there are ideally no restrictions on the relative motions of the particles in the system.

The transmission systems previously described (► Sect. 3.3.1.1. rigid systems) can be combined into forming many different types of mechanisms achieving kinetic effects of different types. This section mentions just a few of them, according to the type of output.

3.3.2.1 Linear Motion

In mathematical terms, motion can be described as a change in position with respect to some frame of reference. Linear motion can therefore simply be defined as the body's final position minus its initial position, defining the distance (*l*). For linear movements, the displacement is conventionally defined relative to right-angle directions: motions proceeding right on the *x*-axis and upward along the *y*-axis are positive, while the opposite directions are negative. Therefore, in mechanics, the sign of a number refers to the direction and not to the magnitude (Knudson 2007). The corresponding vector quantity is displacement (*d*).

Mechanisms achieving linear motion are numerous and have a varied design. Most mechanisms combine elements performing linear motions with others performing rotational motion to produce one or the other as output. Here, a few common mechanisms are mentioned.

Screws are very basic elements that can be used to transform torque into linear motion or vice versa. Screws can be transformed into simple machines, either by rotating the screw shaft through a threaded hole in a stationary object (*stationary nuts with travelling screw*) or by rotating a threaded collar around the stationary screw shaft (*stationary screw with travelling nuts*).

The *slider-crank mechanism* combines a link that can rotate freely around a hinge and a link attached to a roller, restricting it to linear motion (Sclater 2011). The mechanism can convert linear to rotary motion and vice versa. This mechanism is at the base of the linear movement of the pistons, connected with a rod and crankshaft to turn linear motion into rotary motion within a combustion engine (■ Fig. 3.11).

Cam and follower mechanisms are based on a rotating profiled shape (the cam) in direct contact with a lever (the follower) that is moved up and down following

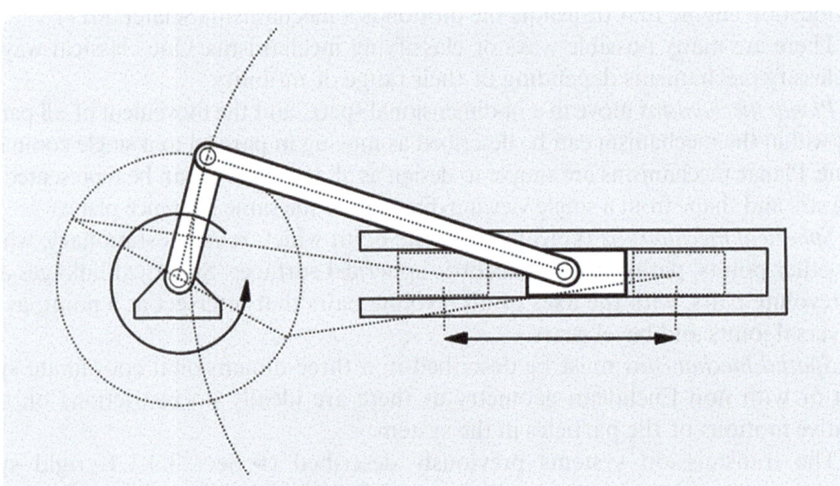

■ **Fig. 3.11** Mechanics of a slider-crank

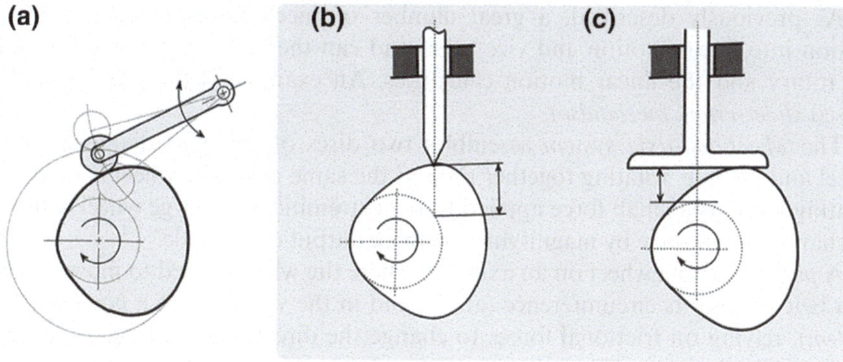

Fig. 3.12 Three plate cams with different types of follower design: **a** an oscillating roller follower; **b** an offset reciprocating knife-edge follower; and **c** a reciprocating flat-face follower. Adapted from Shigley and Uicker (1995)

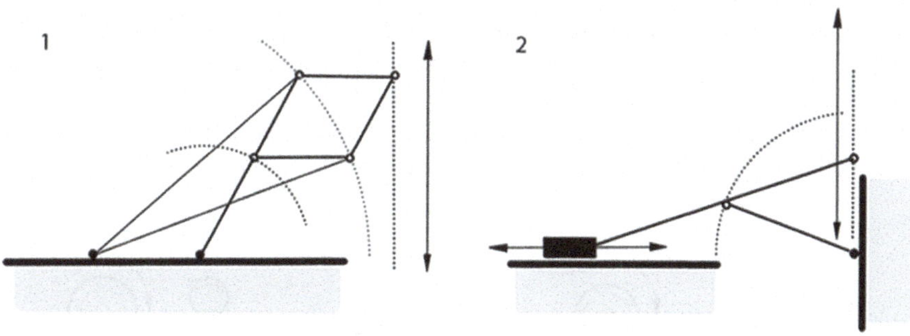

Fig. 3.13 Examples of straight-line generators: (1) a Peaucellier inversor and (2) a Scott–Russell exact straight-line mechanism

the profile of the cam. These mechanisms can generate either an oscillating semi-circular or linear type of motion (Fig. 3.12).

Straight-line generators are specific types of mechanisms able to draw a straight short line of different degrees of precision. Most straight-line generators rely on linkages (mechanisms formed by connecting two or more levers) but can also be achieved using rotary mechanisms (such as the Cardan straight-line mechanism) (Fig. 3.13).

3.3.2.2 Rotational Motion

Angular kinetics explains the causes of rotary motion. Torque (*T*) is the force moment calculated as the product of force (*F*) and the moment arm (*l*) acting on an object, which creates an angular acceleration inversely proportional to the moment of inertia. The overall rotation of the object depends on the balance of torques created by the forces acting on the object (Knudson 2007).

3

As previously described, a great number of mechanisms transform rotary motion into linear motion and vice versa and can therefore be classified in both the rotary and the linear motion categories. An example is the previously discussed *slider-crank mechanism*.

The *wheel-and-axle system* assembles two discs of different diameters, or a wheel and an axle, rotating together around the same axis. The wheel is used as a rotating lever, as a small force applied to the extremities of a large wheel achieves mechanical advantage by magnifying the force output on the axle.

A *pulley* is also a wheel on an axle, but where the wheel is used to move a cable or a belt around its circumference (also found in the variation of a *band and belt system*), relying on frictional forces to change the direction of a force to achieve mechanical advantage. The movements involved shift from linear to rotational and back to linear, with the combination of more pulleys achieving a higher mechanical advantage. By designing the size of the pulleys and the direction of the belt, the speed of rotation can be changed as well as the direction of rotation (Sclater 2011) (◘ Fig. 3.14).

Gears (previously described in ▶ Sect. 3.1.1.4.) are often used in rotating machinery as gear trains (combinations of more gears) to transmit motion in different ways (change direction of rotation, speed, torque, etc.) and achieve mechanical advantage (Sclater 2011).

Spur gears are usually cylindrical with straight teeth parallel to the axis of rotation, which allow to transmit rotary motion between parallel shafts (◘ Fig. 3.15).

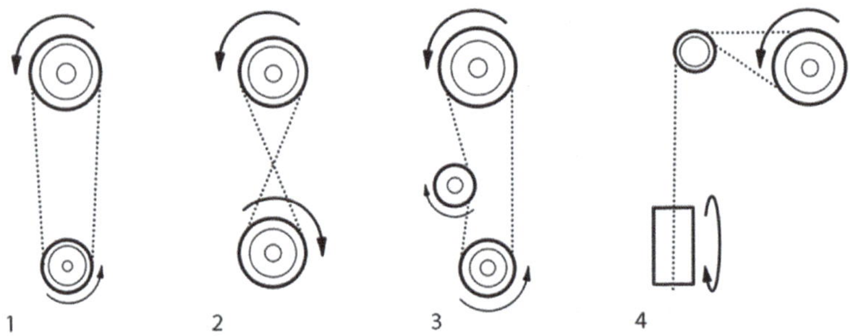

◘ **Fig. 3.14** Band and belt systems in different arrangements: (1) change of speed between the pulleys; (2) change of direction of rotation; (3) combination of change of speed and direction of the rotation in the pulleys; and 4) rotation transmission between pulleys with a different axial orientation

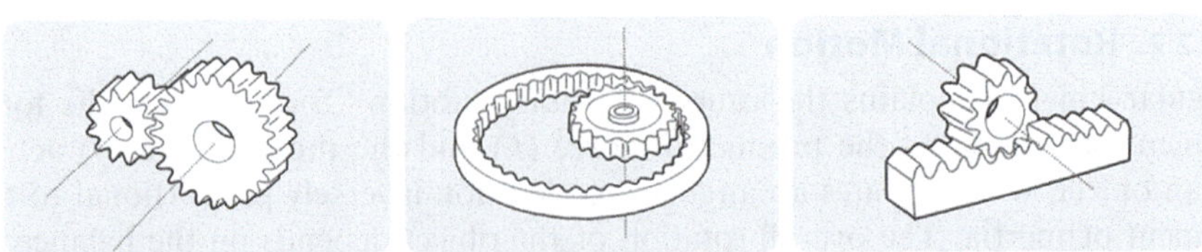

◘ **Fig. 3.15** From left to right, a set of spur gears, an internal gear and a rack and pinion gear

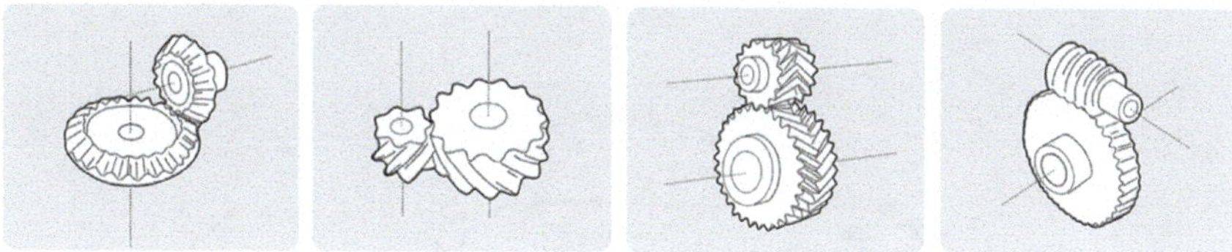

Fig. 3.16 From left to right, a set of bevel gears, helical gears, herringbone gears and a screw gear

Internal or *annual gears* are a variation of the spur gear where one gear is driven inside another gear which has the teeth cut on its inside perimeter (Fig. 3.15).

A *rack and pinion gear* combine a normal gear with a toothed plane. This kind of mechanism converts rotary to linear motion. Funicular railways are based on this mechanism (Fig. 3.15).

Bevel gears are cone-like toothed surfaces transmitting motion between shafts with intersecting axes with any angle (usually 90°) (Fig. 3.16).

Helical gears are spur gears with the teeth cut at an angle, rather than parallelly to the shaft, which allows them to operate more smoothly, with heavier loads and at faster speeds than normal spur gears. *Double helical* or *herringbone gears* have both right- and left-hand teeth cut on the same blank, with the advantage that the thrust forces of the right- and left-hand halves are equal and opposite, cancelling each other out (Fig. 3.16).

Screw or *worm gears* wrap around a screw-like cylindric thread with which they have line contact. These transmit low-speed rotary motion between shafts intersecting usually at a 90° angle (Fig. 3.16).

Four-bar linkages (planar, but also spatial ones) are a common way to trace complex and varied sets of curves. The mechanism is defined by the rigid parts it is made of and the linkages, where the movement is occurring. Tracing the path that each linkage can assume gives a very clear idea of the performed movement (Fig. 3.17).

3.3.2.3 Motion Control

Motion control in mechanisms can be achieved in many different ways, depending on the necessities of the system. Some common aspects that characterize the motion achieved are:

1. Direction of the movement;
2. Speed of the movement;
3. Stability of the movement;
4. Precision of the movement;
5. Timing of the movement;
6. Magnitude of the movement.

Most of these characteristics can be controlled by selecting specific designs of mechanisms, as in the case of simple pulley and belt systems that can be used to change the direction of movement, or speed that can be adjusted using gears

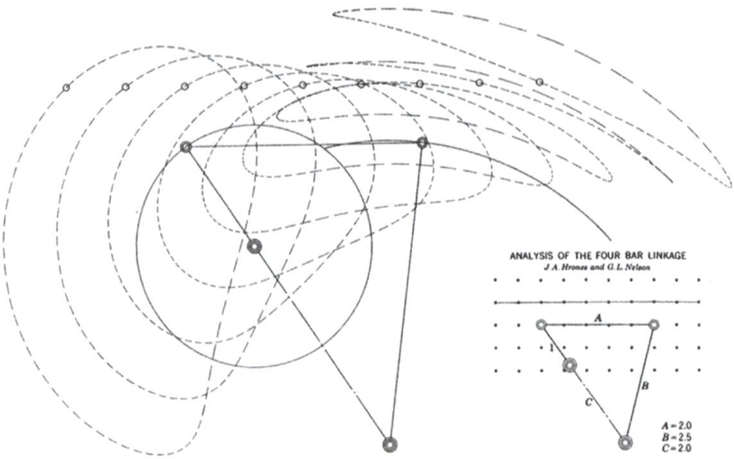

Fig. 3.17 A four-bar linkage which represented all the coupler curves that can be performed by the system (from Shigley and Uicker 1995)

or pulleys of different sizes. As previously mentioned, higher kinetic stability can be achieved using herringbone gears instead of helical gears to control high axial loads within a machinery.

Likewise, high-precision mechanisms are categorized as *indexing mechanisms*, as Geneva wheels (further described in ► Sect. 3.3.2.4), which enable quick and precise changes in position within a mechanism (Shigley and Uicker 1995). Alternatively, small and precise adjustments can be achieved within ranges of micrometres using differential screws and worm gears (as in microscopes or musical tuning mechanisms), or wedges and levers (Shigley and Uicker 1995).

The timing of repetitive actions can be finely tuned using different types of mechanisms that are further discussed in ► Sect. 3.3.2.4.

Mechanisms controlling and mostly restricting the magnitude of motion are also available in many different combinations. A few common mechanisms are:

Detent mechanisms are designed to resist or temporarily keep one body in a specific position relative to that of another. The detent mechanism usually allows to be released by applying enough force to one of the parts (Fig. 3.18).

Clamping mechanisms are used to fasten an object or a surface by applying enough inward pressure upon the object's surface so as to prevent it from moving. Examples are screw clamps used by woodworkers, presses, clip hangers, surgical scissor-shaped clamps, etc.

Latches or *latch release mechanisms* are mechanical fasteners that allow to temporarily and reversibly fix or secure two or more objects or parts (Fig. 3.19). Typical examples are different bolts and hooks used to close doors and windows.

Ratchet mechanisms are used to constrain motion in a mechanism to only one direction, impeding motion in the opposite one, or to transmit intermittent motion (Sclater 2011). A ratchet is usually made of a wheel or a linear rack with

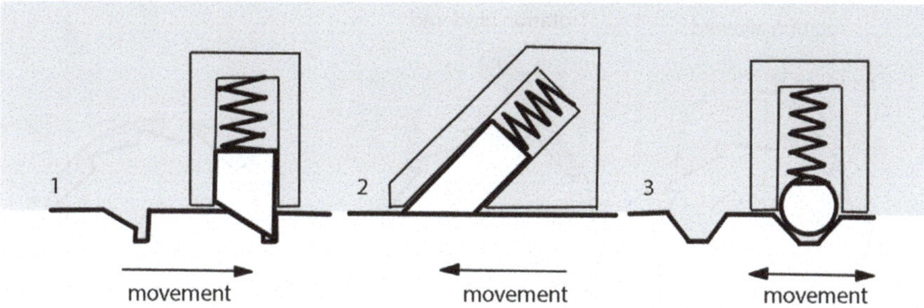

Fig. 3.18 Example of three *detent mechanisms*. Adapted from Sclater (2011): (1) a notch shape detent or a (2) wedge-shaped detent locks the movement in one of two directions; (3) the spherical detent allows constant holding power in both directions

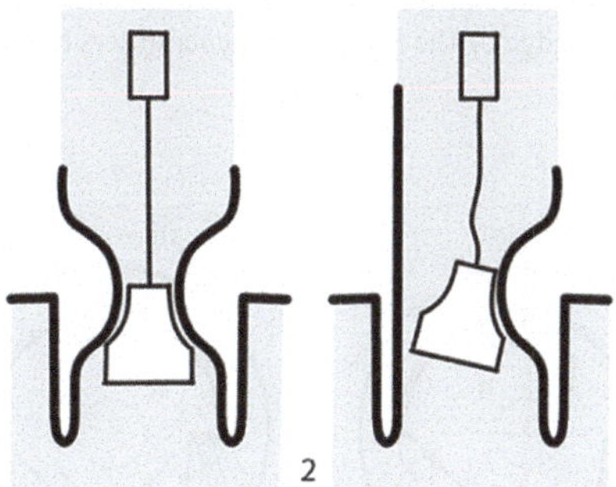

Fig. 3.19 Example of a shape-memory alloy (SMA) *latch*. Adapted from Sclater (2011). (1) The SMA strips have been shaped to secure the knob; (2) as the strips are heated, they straighten out releasing the knob

uniform but asymmetrical teeth, and a rotating spring-loaded stick (the pawl) which engages with only one side of the teeth (Fig. 3.20). Typical applications are in clocks, handcuffs, typewriters, etc.

Clutch mechanisms allow the driving and the driven parts in a mechanism to engage and disengage, controlling power transmission between the two. Clutching mechanisms can be operated externally or internally and have many possible designs (Fig. 3.21).

3.3.2.4 Time-Related Motions

In many cases, a mechanism may need to halt temporarily, because of its specific purpose or while another operation is being performed before it continues its motion. This need has produced many different mechanisms that pause, hesitate or stop and dwell at pre-defined time intervals.

Reciprocating motion is a repetitive up-and-down or back-and-forth linear motion. It is found in a wide range of mechanisms, including pumps and reciprocating engines. For instance, internal combustion engines use the expansion of burning fuel in the cylinders to periodically push the piston down. The linear

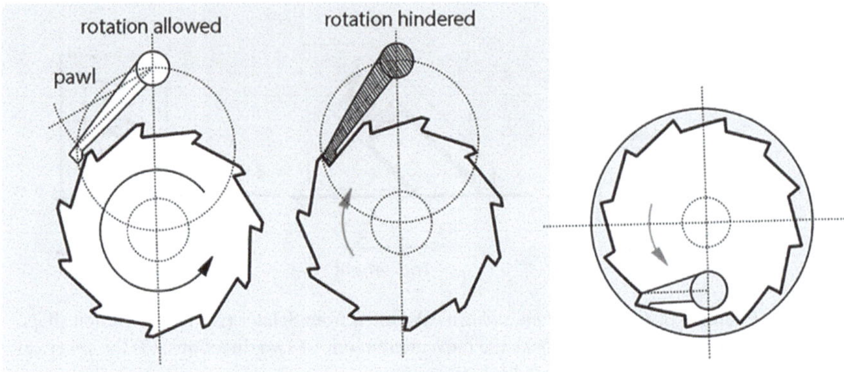

Fig. 3.20 Example of a *ratchet wheel*. Adapted from Sclater (2011). When the wheel turns counterclockwise, the pawls slide over the sloped teeth. When the direction of the rotation is inverted, the pawl will engage with the sharp edge of the first tooth it encounters hindering the rotation. On the right side, an internal ratchet

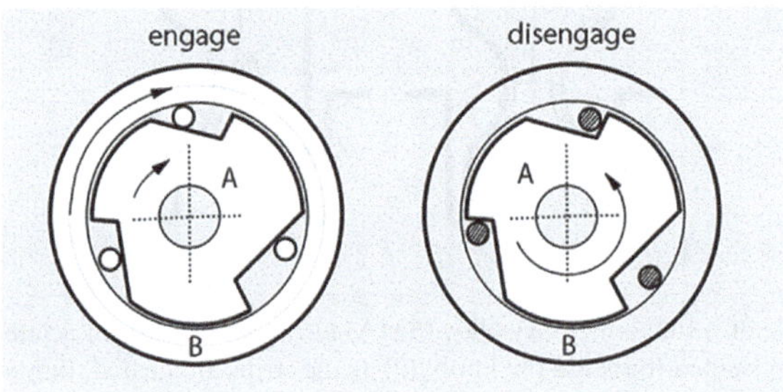

Fig. 3.21 Example of an *overrunning clutch*. Adapted from Sclater (2011). When the driving member (cam A) rotates clockwise, the driven member (ring B) is engaged, as the rollers between the cam and the ring are forced by friction into the wedged spaces between the two (1). However, if the direction of the rotation is inverted, the rollers disengage, and cam A rotates freely inside ring B (2)

movement is repeated over and over, reversing work without the need to stop the mechanism to change direction.

Oscillating motions can be achieved, for instance, with cam and follower mechanisms and can also be used to shape a specific reciprocating motion sequence (Fig. 3.22).

Intermittent motion provides a unidirectional motion occurring during regular or irregular intervals of time. These mechanisms are commonly used in clockworks and movie projectors.

Examples are *Geneva drives* that have a rotating mechanism with a pin that in specific intervals of time reaches into specifically positioned slots in the driven wheel. The slots can be designed with in different numbers and variants of star shapes, the most common one being the Maltese cross (Fig. 3.23).

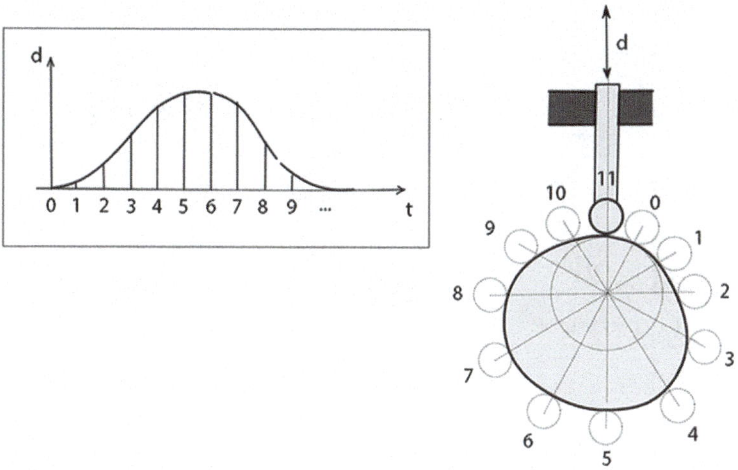

Fig. 3.22 Design of the oscillation in a cam and follower mechanism. The graph to the left describes the lift achieved by the follower (*d*) in relation to the position of the follower in the different stations (*t*). To the right, development of the cam profile according to the graph: the cam is held stationary, while the follower is rotated through the stations 0–11. Adapted from Shigley and Uicker (1995)

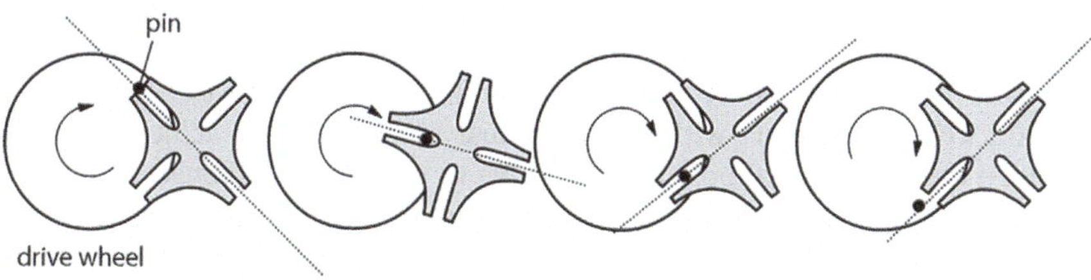

Fig. 3.23 Geneva drive (or Maltese cross) mechanism. The rotating drive wheel engages the driven Maltese cross for a brief moment, using a pin that turns it by 90°

The rotating drive wheel is usually equipped with a pin that reaches into a slot located in the other wheel (driven wheel) that advances it by one step at a time. The drive wheel also has an elevated circular blocking disc that "locks" the rotating driven wheel in position between steps.

Another example is *escapement mechanisms* that provide an even series of stop motion sequences, combining the mechanics of a ratchet system with that of an intermittent mechanism. In this case, the toothed component is rotated in only one direction, while the pawl intermittently restricts and releases the gear train, allowing it to move forward with rhythmic sequence. Escapement mechanisms are therefore commonly used in mechanical timekeeping devices from pendulum clocks to portable watches (Fig. 3.24).

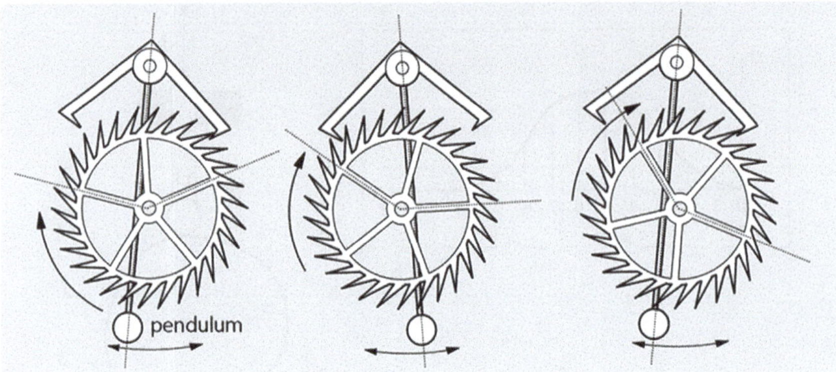

Fig. 3.24 Example of an *anchor escapement* used in pendulum clocks. Energy is transferred from a pawl–pendulum to the gear train which releases up to a certain time the toothed wheel, enabling it to turn at regular intervals

3.4 Antagonist Mechanisms

In a biomechanical context, most mechanisms described in ► Sect. 3.3.1.1. Rigid Systems and ► Sect. 3.3.1.2. Soft-Bodied Systems cannot produce both movement and countermovement, so force must be transmitted so as to return the mechanism to its initial state (Biewener 2007). Therefore, most agonist mechanisms are working in pairs with an antagonist mechanism, to achieve actuation and return to their original position to be able to repeat the sequence.

In man-made mechanical systems however, many mechanisms do not need an antagonist. Specifically, most mechanisms based on rotation movements are able to perform self-reversing actions. So, since the mechanism is so efficient, why has the wheel not evolved in nature? There are different theories to that.

A first theory states that the wheel-and-axle system is an "all-or-nothing" structure and very sensitive to imperfections (not perfectly round, size of the axle, connection, etc.) and could therefore not have been developed by successive attempts and in phases (Anthony 2010).

Another plausible reason is that the wheel needs a frictionless connection around its axis. To allow infinite rotation of the wheel in one or in the other direction, the wheel cannot be physically attached to a body. In nature however, an organism needs to be attached to its appendages—to control movement, growth and healing and to provide a constant supply of nutrients. This also creates a problem of how to grow the wheel in the first place, as well as for its operation through muscles.

Finally, wheels have been recognized to be highly impractical means of locomotion on uneven terrains, where legs and specifically adapted foot variations are much more efficient (Biewener 2007).

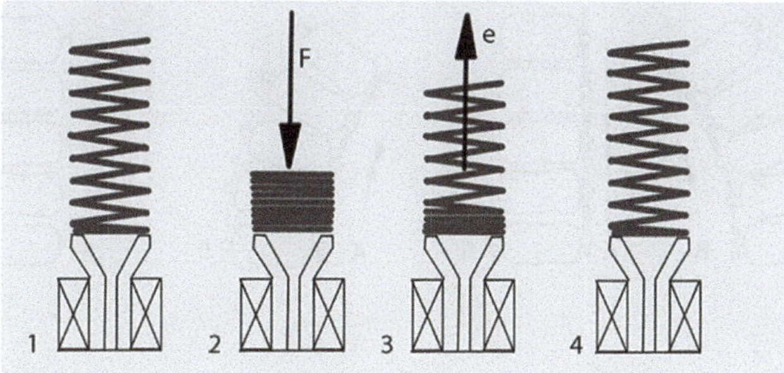

Fig. 3.25 Simple mechanical scheme of a spring element in different stages: (1) the spring is in its relaxed state, at full length; (2) a force (F) is applied to the system, compressing the spring which accumulates elastic energy (*e*); (3)–(4) as the force is removed, the spring regains its initial shape, giving back the elastic energy accumulated in step 2)

In this context, common strategies from the fields of mechanics and biomechanics that can be used to provide antagonist movement are discussed. These are mainly of two types: spring and muscular antagonisms.

3.4.1 Spring Antagonism

Spring antagonism relies on extremely common and simple logics. A spring element—any kind of material or structure with elastic proprieties—stores energy from the agonist movement as the forces acting produce work on the body deforming it. As the agonist force(s) are removed, the body returns the energy contained in the scaled or otherwise deformed (bent, buckled, torqued, etc.) shape to its initial pre-stressed state of equilibrium.

The capacity to act as antagonist depends on the shape and substance of the material, which also defines the range and the timeframe of stress within which the elastic proprieties are kept, allowing to fully reverse the movement.

Spring antagonism is versatile and is therefore very diffused in the mechanical and the natural contexts (Fig. 3.25). In nature, it is often found to work as single antagonist or in combination with other antagonist mechanisms in plants and in animals (Persiani 2018).

3.4.2 Muscular Antagonism

Muscular antagonism is activated in the same way as muscular agonism and requires therefore a complex body structure and coordination. The combination of muscular agonism and antagonism is very common in the animal kingdom and lies at the base of motion in all vertebrates (Fig. 3.26).

3

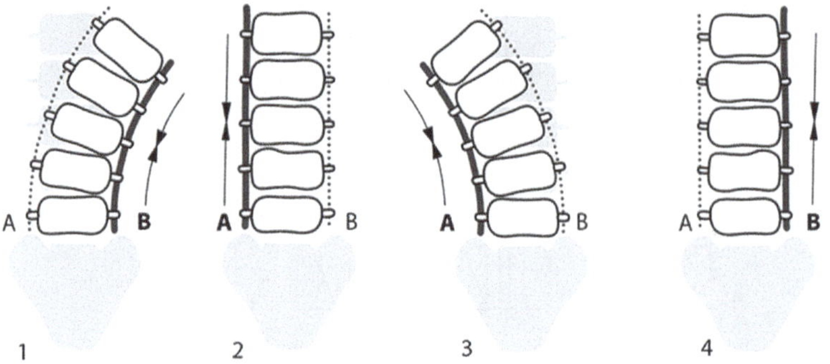

Fig. 3.26 A simplified scheme of muscular antagonism in a vertebrate column. Muscle A and muscle B work in pairs, opposing each other. When muscle B is contracted, muscle A is stretched (1, 4), and when muscle A is contracted, muscle B is stretched (2, 3)

Apart from reversing a specific movement, agonist-antagonist muscle combinations can perform other variations of motions among which:

- Complex and precise movements, as in composite muscles as tongues where the contraction of multiple muscle bundles is combined to perform a broad series of actions, making the distinction between agonist and antagonist difficult (Persiani 2018);
- Stiffening of joints, achieved through the simultaneous activation of agonist and antagonist resulting in a stiffening of the body structure.

3.5 Amplification Mechanisms

The purpose with most mechanisms is not only to transmit motion, but also to achieve mechanical advantage, or a force amplification, when forces are transmitted from one body to another. The system preserves the input power and trades off forces against movement to obtain an amplification in the output force.

In nature, power amplification mechanisms often work in combination with other systems, achieving mechanical advantage through the interaction on different levels of different systems.

In mechanics, the mechanical advantage is calculated as the ratio of output force to input force in the system, assuming there is no friction. It applies to forces in simple machines like levers, inclined planes, wedges, wheel and axles, pulley systems or jackscrews.

These simple machines have already been discussed in the previous part ▸ Sect. 3.3.1.1. Rigid Systems. In this section, further amplification mechanisms are discussed. Most of them can be led back to mechanical principles based on lever systems. The aim is however to highlight specific potentials and combinations of specific systems.

3.5.1 Elastic Amplification

Elastic amplification makes use of the flexible proprieties of selected parts to store mechanical energy and function as power amplifiers. When deformed, stretched or compressed, a body with elastic proprieties is loaded with potential energy by the forces acting on it. As the forces or constraints are removed, the body returns to its initial, undeformed, resting position—giving back the potential energy in the opposite direction to which it has been loaded (Viegas 2005).

The amplification effect mostly relies on the acceleration as the elastic body returns to its state of equilibrium. As previously seen in part ▶ Sect. 3.1.2. Force-Time Principle, the magnitude of a force strongly depends on the acceleration imposed on it. In this sense, elastic energy is more significant as contributor to the actuating efficiency than as actuating force.

In systems combining elastic power amplification with lever principles, the action's performance is improved by the principle of a catapult as the stored energy is released very rapidly in the final stages before the action. The energy returned in the recoil equals the work done to deform the body but is returned at a greatly increased rate.

Natural elasticity refers to the flexible and elastic material proprieties of a body. Typical examples are rubber bands, rubber pads, etc. These materials are strategically placed to provide maximized deformation with minimal effort and easy recovery of energy, combined with good movement control. In these cases, the type of effort applied is typically a normal force, either compressing or stretching the body. Body parts with natural elastic proprieties are commonly used in animal biomechanics. This is the case of clams, where the closing of the shell compresses a flexible hinge ligament located at the hinge connecting the shells. As the contracting muscle keeping the shells closed relaxes, the ligament recoils elastically opening up the shell (Persiani 2018).

Surface tension uses the elastic proprieties of a membrane to transmit motion or load a mechanism. When the constraining factor that deforms the membrane is removed, the system snaps back to the original shape. In this case, the force can be applied either parallelly to the membrane's surface, stretching it, or horizontally to the plane, bending it. For example, in most fishes, the elasticity of the outer skin contributes to transmit the swimming motion along the body (◘ Fig. 3.27).

The elastic material proprieties are often present in combination with other amplification mechanisms, such as with unbalanced systems and geometric features using lever systems.

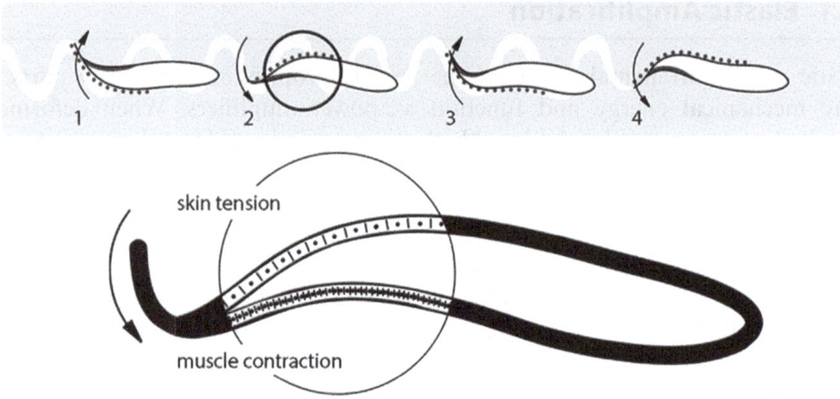

Fig. 3.27 Over, schematic representation of a fish swimming (1–4), the muscle contractions (bold) and the tension of the elastic skin (dotted). Under, focus showing how the elastic skin is tensed on the left side of the fish while a muscle contraction happens on the right side. When the right muscle relaxes, the skin on the left side contributes to the countermovement, returning kinetic energy into the opposite direction, amplifying the effect of the upcoming muscle contraction on the left side

3.5.2 Instability Amplification

Unbalanced systems typically make use of the elastic proprieties of some of their mechanical parts to bring the mechanism to its static limits. This usually requires a preparation time during which the system is loaded with elastic energy. When the constraining factor is removed, suddenly breaking the balance, the system snaps back to its relaxed shape. The sudden release and acceleration of the concentrated forces are used to release a stronger force output and a higher velocity.

Instability amplification is broadly used in plants for seed dispersal and can be subdivided into snap-buckling and fracture-dominated systems.

Snap-buckling systems switch between two opposite and complementary geometries. This is, for instance, the case of the Venus flytrap, a carnivorous plant which is known for being able to close its leaves at a speed high enough to trap its prey. It uses a combination of surface tension and snap-buckling. The convex structure of the leaves is stretched out and tensed to its geometrical limit where an extremely small force can break the balance and instantly flip the leaf from its convex into a concave shape (Persiani 2018). In this case, the amplification effect is used not so much to provide a higher force output but to achieve high speed to trap the prey. Snap-buckling geometries can also be found in man-made contexts where concave or convex shapes are realized with thin elastic and self-supporting surfaces (Fig. 3.28).

Fracture-dominated systems are usually the ones allowing to achieve higher speeds but have the disadvantage that the mechanism is not reversible. In these systems, the energy that is built up within the mechanism is released by tearing thin structural elements when the planned pressure is reached. The Himalayan balsam plant uses this kind of system in its pods for seed dispersal. The seed pods

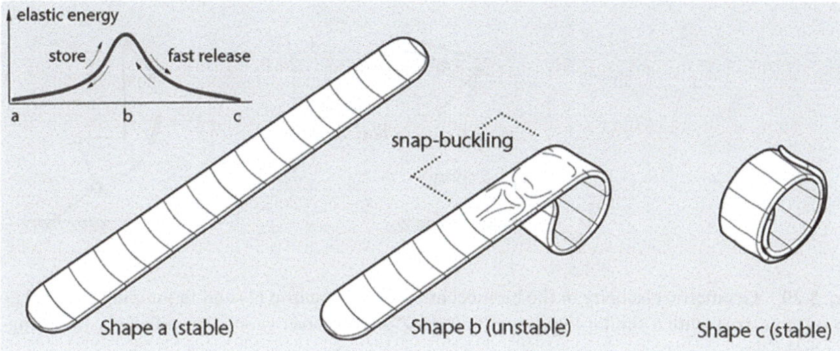

Fig. 3.28 Snap-buckling in a metallic ruler bracelet. This type of gadget switches between a stable linear concave shape (**a**) that can be used as a ruler, and the opposite stable rolled-up shape (**c**). When the first shape (**a**) is bent and rolled, elastic energy is stored in the geometry, until it reaches its limit of stability and the energy is quickly released, popping the geometry into its second stable shape (**c**)

combine a pentaradial elastic geometry with a fracture-dominated trigger. Five curled elastic valves are straightened out and connected to one another by a thin tissue. As the tissue dries up, the elastic curling in the valves drives the simultaneous cracking of all the connecting tissues, the instantaneous coiling of the seedpods and the propulsion of the seeds (Persiani 2018).

3.5.3 Geometric Amplification

In geometric amplification, adaptable structures are designed to switch between multiple states of equilibrium allowing flexibility in these structures to deform or react in different ways depending on boundary, load and geometrical configurations (Menges and Hensel 2008).

Geometric elasticity. Partially elastic features can be integrated into semi-rigid elements and systems made of semi-rigid elements through the design of surface and structural flexibility to deformation. These proprieties are generally used to control high strains through geometrical energy dissipation. Geometries that are typically used are circular, helical, curved and bent *V*- and *Z*-geometries (Vincent 2008; Biewener 2007) (Fig. 3.29). Geometric elasticity can be found in biomechanical systems on various scales, from the typical Euler buckling in human bones to the biomechanics of running where the deformation of leg acts as a spring element, absorbing and returning part of the kinetic energy.

Indirect actuation is typically used in organisms built with rigid elements relying on the use of advanced lever systems. In basic lever-pivot systems, the power output is proportional to the direct force of the muscle (Nachtigall 2005) and speed (frequency of the effort) is limited. When the lever-pivot system is more complex, it allows higher speed and force than the direct muscle's capacity. When rowing, the simple lever system of the human arm is further amplified by add-

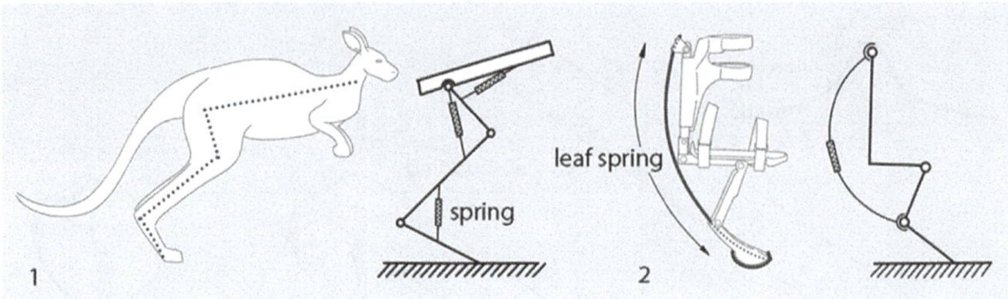

Fig. 3.29 Geometric elasticity in the biomechanics of a kangaroo (1) and in jumping stilts (2). Both systems work with a similar logic, combining a Z-shaped lever geometry and an elastic spring-like antagonism

ing the oars as lever. Although the muscular input force is comparable, the force exerted at the end of the oar will be much higher compared to the force exerted at the end of the hand (Persiani 2018).

Auxetic geometries allow geometric amplification in the extent that linear actuation of the geometry can be transformed into a bi-dimensional expansion or shrinking of the geometry (see also ▶ Sect. 4.3.1. Auxetic Materials).

3.6 Control System

Within a system in motion, the components must coordinate and work together seamlessly to perform their assigned functions. In many cases, the information that needs to be processed to achieve efficiency (in mechanical systems) and survival (in biological systems) is complex and manifold, requiring one or more body parts to act as regulator.

The control system is broadly intended as the structure processing the information, receiving, integrating and interpreting incoming information and coordinating response by signalling subordinate parts on how to act. The capacity of a control centre can be described in terms of:

- Amount of incoming information that can be handled;
- Amount of detail in the information;
- Amount of possible combinations of solutions;
- Response time.

The more complex the information handled, the more energy-intensive the control centre. As energy efficiency is at the end what every system must strive for, in terms of costs versus relevance of the action, highly complex processing systems might provide great output, but at a very high cost. So, how can good coordination and control be achieved, while using as little resources as possible?

Fig. 3.30 Progressive emergence of a Turing pattern

3.6.1 Self-organization

Self-organization is an overarching concept that can be applied in a great variety of contexts. It implies the ability of a system to develop and sustain its own form of order arising from the local interconnections between its parts, without any deliberate control from outside. The capacity to rapidly adapt to change and regenerate itself is typically managed through feedback loops. This chaotic unpredictability of the overall behaviour is strongly influenced by exchange of energy, matter and information with the environment, which break the existing symmetries while forming new patterns (Gruber 2011).

Self-organization is found in both animate systems, as in phenomena such as homeostasis, morphogenesis and motion coordination, and inanimate systems from market economy (Krugman 1995) to collective intelligence in media, such as the Wikipedia platform (Deneubourg et al. 2003).

Turing patterns are naturally occurring patterns such as stripes, spots and spirals that arise out of a homogeneous uniform state (Turing 1952), as patterns on animal coats or sand formations due to the action of the wind (Fig. 3.30).

Self-assembly refers to a self-organization process in which a disordered system of components assembles through local interactions to form an organized and pre-designed structure or pattern, without any external coordination. In a design context, self-assembly can bring several advantages to a solution, from easier transport to assembly on distance—in places difficult (hostile and extreme environments such as space) or too small to reach (as in the case of microrobotics for medical applications) (Fig. 3.31).

3.6.2 Self-amplification

Self-amplification is a strategy that allows a system with centralized processing to share basic information with peripheral centres. Initiation of the command depends on the central processor, and the communication with the subordinate agents however depends on a chain reaction designed to occur along a pre-defined intentional pattern, animating a wider area.

3

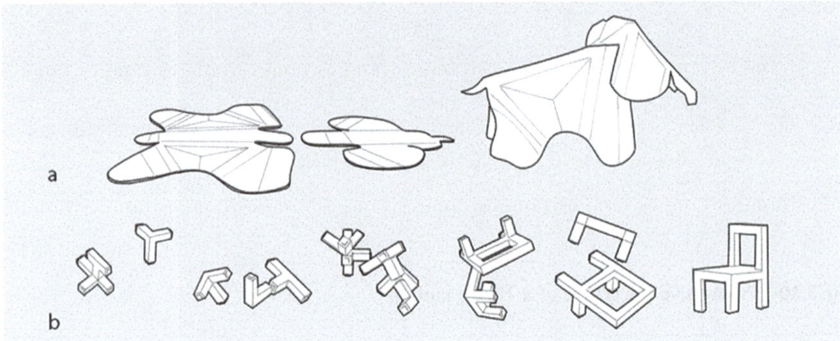

Fig. 3.31 Two different self-assembly designs developed by the MIT self-assembly laboratory (MIT Self-Assembly Lab 2020): **a** self-assembly relying on the activation of programmed materials, bending along pre-designed folding lines, and **b** self-assembly relying on a cumulative selection process within a turbulent fluid. The single parts are equipped with unique couplings defining their precise location within the structure, which eventually assemble being repetitively tossed around in the turbulence

All the elements in this system are connected to each other by an intentional or casual pattern. Action of one element causes reactions in the neighbouring elements, being transmitted along successive stages of the chain and resulting in a spatial amplification of an effect as a result of stored potential energy.

The design of the connections can be very simple and relatively inexpensive. The pattern of these connections impacts the speed, the order of succession of the actions and ultimately the time lapse between the action and reaction of the first and the last unit in the chain, depending on its geometry (Fig. 3.32).

This kind of system is relatively simple to design and to control but relies heavily on the connections between few components. The problem in the chain reaction is that if a connection eventually fails, the whole system risks to be fully or partially put out of operation. Also, the chain reaction system might not be

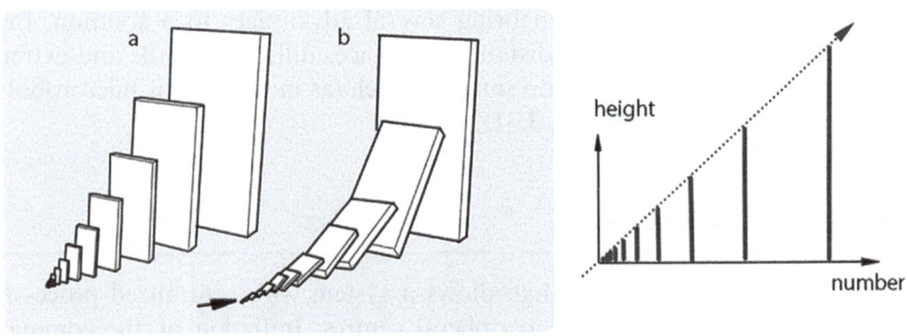

Fig. 3.32 A self-amplifying domino effect using a series of 13 dominoes (**a**, **b**), where each successive domino is 1.5 times the size of the one knocking it over (scheme on the right). As the single domino falls, it adds gravitational kinetic energy to its mass, resulting in a step-by-step force amplification that topples the bigger dominoes (van Leeuwen 2013)

extremely quick to react to exterior input or change for two reasons: (1) reaction must happen in a specific unit of the chain; (2) reaction speed depends on the pattern of connection(s), hence on the time needed for the domino effect to the last element in the reaction chain.

In this type of system, energy use is optimized by building a system of selected parts specialized on specific tasks:

- The amount of incoming information is selected depending on the task.
- The amount of detail is only processed if necessary.
- The amount of possible combinations of solutions are reduced.
- Response time is adapted, depending on the urgency of the task.

3.6.3 Swarm Intelligence

Decentralized control works through networks made out of low-intelligence agents able to locally answer to basic tasks in an individual manner. It allows to minimize the amount and complexity of the information handled as minor control centres are designed to intercept routine information and to deal locally with simple tasks related to it, saving energy. Connecting the minor centres through simple sets of rules and principles of collective behaviour and self-organization enables them to react in unison as one macro-system, minimizing the effort spent on coordination.

Swarm intelligence systems are typically made of populations of simple agents interacting locally with one another and with their environment (Fox and Kemp 2009). When the groups of individuals are sufficiently large, the combinations of interactions can develop emergent behaviours that are very difficult to predict. As in other decentralized control networks, swarm intelligence relies on minor centres operating under simple rules of interaction allowing to limit the amount and complexity of incoming information, enabling reaction on short timeframes while allowing to generate complex combinations of solutions over a population of individuals.

Swarm intelligence can therefore be typically found in large groups of individuals as in animal swarms, shoals and flocks. It allows the group to coordinate and quickly react to recurring issues as direction of movement, defence mechanisms, migration time and foraging. For example, for specific species of fish, shoaling is thought to increase individual security with lower chances of predation and higher chances of foraging, as well as hydrodynamic capacity (Alexander 2004).

Typical main parameters used to describe these large systems are:

- Size, given by the number of individuals (often in thousands or millions);
- Density, given by the number of individuals divided by the volume occupied—although density is not necessarily constant throughout the group;
- Polarity, describing the extent to which the individuals are pointing to the same direction.

The coordination of movements seems to be controlled by a distancing/cohesion rule—regulating the acceptable distance range between individuals avoiding collisions, crowding or distancing losing group coordination—and the alignment rule—allowing individuals to steer towards the same direction (Hemelrijk and Hildenbrandt 2011). Group decisions are taken through *quorum sensing*, an auto-inductive decision-making process based on the sensed changes in population density: once a minimum number of individuals in a group perform a specific action, the probability of other group members performing the same action steeply increases, resulting in a quorum response (Ward et al. 2012).

3.6.4 Redundancy

Redundancy is a design technique assuring security in the functioning of the system by giving the possibility to perform the same action in many different ways. In case of failure or malfunctioning of one part of the system, the action is not compromised but can be performed anyway. This strategy is used in multiple contexts both in nature and in technology.

In engineering, the concept of redundancy is mainly applied in specific key mechanisms, as to ensure the reliability of a security system: controls are duplicated, eventually triplicated, to prevent a single device failure from disrupting the whole mechanism or network.

In nature, redundancy seems to be widespread in the genomes of higher organisms: two or more genes perform the same function, meaning that the inactivation of one of these genes has little or no effect on the outcome of the biological phenotype (Nowak et al. 1997).

3.6.5 Design of Failure

Man-made structures are designed to have no life after failure. Broken objects and mechanisms are simply thrown away and replaced with new ones. As the possibility of the structure to heal and change in time is never taken into consideration, the safety factor ratio "*s*" (load producing failure/greatest load expected in use) used in man-made structures is extremely high.

Natural structures on the other hand are designed for durability and survival, not for high performance. While human engineering mostly prioritizes the maximization of the output (speed, size, force, etc.), nature gives up absolute strength to prioritize lightness and achieve a better energy balance, accepting fracture when a certain load limit is exceeded (Gruber 2011). The safety factor is very low, and structures are characterized by imperfections, asymmetry, irregularities and weak points. Although imperfection might be difficult to manage, it can produce more durable structures if dealt with in the right way. The careful design of areas of weakness in a structure can help protecting important parts of the structure (Menges and Hensel 2008).

Vertebrate bones, for instance, are at the same time rigid enough to allow mechanical advantage using them as lever systems, and flexible allowing shock absorption. The bones are not straight, as would be expected in a man-made system, but present a bended shape (the "Euler buckling") that allows the bone to partially bend dissipating overloads during movement. The force resisted is less, but the energy absorbed is greater.

3.7 Conclusion

This chapter has reviewed a series of aspects that are central to an efficient energy and motion transmission. The mechanical parameters have been broken down into simple principles to facilitate their integration in the early stages of the conceptual design. More specifically, the reviewed aspects include:

- Core principles of biomechanics on which to base the kinetic design of the autoreactivity;
- Types of actuating systems, which impact, depending on the source of energy employed, whether the device qualify or not as autoreactive;
- Types of agonist mechanisms, bringing to the selection of the mechanism of transmission to the type of movements performed;
- Types of antagonist mechanism (when required);
- Possible ways to amplify the force or the range of motion;
- Additional control systems to aid coordination between parts and simplification of information processing.

As a brief inventory of conceptual tools, these single aspects need to be combined and skilfully matched with the aspects described in the other chapters in order to achieve smart and efficient autoreactive designs.

References

Alexander RMN (2000) Storage and release of elastic energy. In: Nigg BM, MacIntosh BR, Mester J (eds) Biomechanics and biology of movement. Human Kinetics, Champaign, Ill

Alexander R McN (2004) Hitching a lift hydrodynamically—in swimming, flying and cycling. J Biol 3(2):7. ► https://doi.org/10.1186/jbiol5

Anthony DW (2010) The horse, the wheel, and language: how bronze-age riders from the eurasian steppes shaped the modern world. Princeton University Press, ISBN, p 9780691148182

Biewener AA (2007) Animal locomotion. Oxford University Press, Oxford, UK. ISBN 978-0-19-850022-3

Burrows M, Sutton G (2013) Interacting gears synchronize propulsive leg movements in a jumping insect. Science 341(6151):1254–1256. ► https://doi.org/10.1126/science.1240284

Deneubourg C, Sneyd F, Bonabeau T (2003) Self-organization in biological systems. Princeton University Press. ISBN 0-691-11624-5

Encyclopædia Britannica (2012) Muscle, linkage. ► http://www.britannica.com/. Accessed on 16 Nov 2012

Fox MA, Kemp M (2009) Interactive architecture. Princeton Architectural Press, New York

Gruber P (2011) Biomimetics in architecture, architecture of life and buildings. Springer, Wien. ISBN 978-3-7091-0331-9

Harrison D, Comyns T (1997) Biomechanics of the sprint start. Sports Med 23(1):11–20. ► https://doi.org/10.2165/00007256-199723010-00002

Hemelrijk CK, Hildenbrandt H (2011) Some causes of the variable shape of flocks of birds. PLoS ONE 6(8):e22479. ▶ https://doi.org/10.1371/journal.pone.0022479

Kier WR (2012) The diversity of hydrostatic skeletons. J Exp Biol 215:1247–1257. ▶ http://jeb.biologists.org/content/215/8/1247.full

Knudson D (2007) Fundamentals of biomechanics. Springer, New York

Krugman P (1995) The self organizing economy. Blackwell Publishers. ISBN 1557866996

Menges A, Hensel M (2008) Morpho-ecologies. Architectural Association, London

MIT Self-Assembly Lab. (2020) Fluid-assembly chair. Retrieved 12 March 2020 from ▶ https://selfassemblylab.mit.edu/fluid-assembly-chair

Nachtigall W (2005) Biologisches design, Systematischer Katalog für bionisches Gestalten. Springer, Berlin

Norberg UM (1981) Flight, morphology and the ecological niche in some birds and bats. In: Day MH (ed) Vertebrate locomotion, symposia of the Zoological Society of London, no 48. Academic Press Inc., London, UK. ISBN 0-12-613348-4

Nowak MA, Boerlijst MC, Cooke J, Maynard Smith J (1997) Evolution of genetic redundancy. Nature 388(6638):167–171. ▶ https://doi.org/10.1038/40618

Persiani SGL (2018) Biomimetics of motion, nature-inspired parameters and schemes for kinetic design. Springer, Cham, Switzerland. ▶ https://doi.org/10.1007/978-3-319-93079-4

Petit F, Chalon M, Friedl W, Grebenstein M Albu-Schäffer A, Hirzinger G (2010) Bidirectional antagonistic variable stiffness actuation: analysis, design & implementation. In: Proceedings of the 2010 IEEE international conference on robotics and automation anchorage convention district, Anchorage, Alaska, USA, 3–8 May 2010. ▶ https://doi.org/10.1109/robot.2010.5509267

Sclater N (2011) Mechanisms and mechanical devices sourcebook. McGraw-Hill. ISBN 978-0-07-170442-7

Shigley JE, Uicker JJ Jr (1995) Theory of machines and mechanisms, 1st edn. 1980, 2nd edn. McGraw-Hill Inc., USA

Turing AM (1952) The chemical basis of morphogenesis. Philos Trans Roy Soc Lond B 237(641):37–72. ▶ https://doi.org/10.1098/rstb.1952.0012

van Leeuwen JMJ (2013) Domino magnification. Cornell University, Physics, Popular Physics. ▶ arXiv:1301.0615

Viegas J (2005) Kinetic and potential energy: understanding changes within physical systems. The Rosen Publishing Group Inc., New York. ISBN 1-4042-0333-8

Vincent J (2008) Imperfection as design parameter. In: Hensel M, Menges A (eds) Morpho-ecologies. Architectural Association, London

Ward AJ, Krause J, Sumpter DJ (2012) Quorum decision-making in foraging fish shoals. Public Library Sci One (PLoS) 7(3). ▶ https://doi.org/10.1371/journal.pone.0032411

Energy of Autoreaction

Contents

S. G. L. Persiani, *Design of Autoreaction*,
Design Science and Innovation, https://doi.org/10.1007/978-981-15-6178-8_4

Energy is described as the physical quantity that measures the ability of a body or a physical system to perform work. In physics, a force acting upon an object and causing it to displace is said to do work upon the object (Viegas 2005): work (*W*) equals force (***F***) times distance (***d***) moved.

$$W = \boldsymbol{F} \cdot \boldsymbol{d}$$

If an object is capable of doing work, it has "stored work", which is known as energy. Hence, energy and work have the same unit of measure, the Joule (J).

4

In other words, to achieve change in an object, a transfer of energy must take place. The law of conservation of energy states that energy in an isolated system is constant, which means that there is a fixed amount of energy in the entire universe, which can never change. Energy cannot be created nor destroyed, but changes from one form to another (Viegas 2005).

The environment is a complex synergetic system acting on and interacting with given subordinated synergetic systems (ecosystems, organisms, etc.) by exchanging different types of energy, determining each other's development. Subordinated synergetic systems are highly dependent on their environment, with which they must constantly interact in order to sustain and develop themselves. While superior synergetic systems mould and define their subordinate systems, the latter also interact with each other influencing the superior system, in some cases actively redefining it. For instance, a subordinated system (mankind) impacts in a domino-like effect its superior synergetic system (the planet), forcing other subordinated sibling systems (other species) to adapt to the new conditions, or disappear. In 1986, when a unit of the nuclear power station plant of Chernobyl exploded releasing radioactive material into the environment, man jeopardized the balance of its own environment, and was forced to permanently evacuate a 30 km area around the plant due to the unsustainable levels of radioactivity. The effects of the accident had, however, a much broader impact, modifying permanently the energetic balance of that specific ecosystem. On one hand, it made the living conditions of its sub-systems impossible, causing mass-destruction, on the other, it opened up new ecological niches for evolution to take advantage of. A study published in recent years shows how some birds may be adapting to the levels of radiation (Galvàn et al. 2014; Zielinski 2014) even obtaining beneficial hormetic[1] effects. Those specific subordinate systems adapt to fit the new energetic conditions of their superior system.

Interaction between physical systems on all scales and in all domains—superior synergetic systems, subordinate systems, ecosystems, organisms, buildings, etc.—can be described in terms of exchange of different kinds of energy. For a sub-system to be considered sustainable, the energy exchange between this system, its superior and any other sub-system it comes into contact with must be

---

1 Hormetic effects are an adaptive responsivity of the organisms to a moderate and (usually) intermittent physical stress (Mattson 2007).

constant. This is typically the case of an ecosystem: from an energetic point of view, it is a balanced system of constant energetic exchange between its parts—whether this energy exchange happens as a predator feeds on a prey or as organisms enter a mutual symbiose.

To a certain amount, part of the energy contained in a system is unavailable for being converted into work and is instead lost to dissipation or friction. This tendency is called "entropy". This means that it is impossible for a man-made machine to convert all input energy into a useful energy output and achieve 100% efficiency. When powering a lightbulb with electrical energy, the ordered form of energy which is expected is in the form of light. The transformation of electricity into light, however, generates an energetic by-product in the form of waste heat. In the context of autoreactive systems, this kind of "leftover energy" becomes an interesting opportunity for multi-purpose design.

This chapter explores the sources of "waste energy" and other forces that can be taken into consideration to achieve kinetic motion—atmospheric agents (wind, rain, hail, etc.), gravity, electromagnetism and chemical energy, etc.—analyses, categorizes, maps and suggests possible applications to achieve, implement and control autoreaction.

4.1 Energy and Autoreactivity

Energy conservation and optimization are key factors in all living organisms: as part of the ecosystem, our buildings and artefacts also need to balance the energetic exchange with their surroundings. Autoreactive systems divert available waste energy deriving from other processes in the surrounding environment and transform it into kinetic energy to perform the set purpose, adapting along with change in their superior synergetic system. Residual energy sources are conveyed and repurposed, embedding more coexisting functions into one single system.

Substances and objects may have the capacity to store and release energy, becoming potential sources, but are not energy themselves. The designer of an autoreactive system needs to be aware of the nature and timing with which energy flows will be available, in order to decide on the kinetic output and effectively control it. While the needs of man-made devices change at an hourly, diurnal and seasonal rate, the natural energy sources in our environment change at a totally different frequency (Harris and Wiggington 2002). The human performance drivers are typically characterized by unpredictability and instability giving rise to potential conflicts between the availability of the energy source and the timing of the functional needs. What more is, specific functions required by the designed devices might not share any common parameters with the energy sources that are available at that specific moment, complicating the possibility to design the reaction. Therefore, understanding the underlying relationships between factors as the energetic flows and the devices' functionalities, that are the main elements in the autoreactive design equation, is of primary importance for architects and engineers (◘ Fig. 4.1).

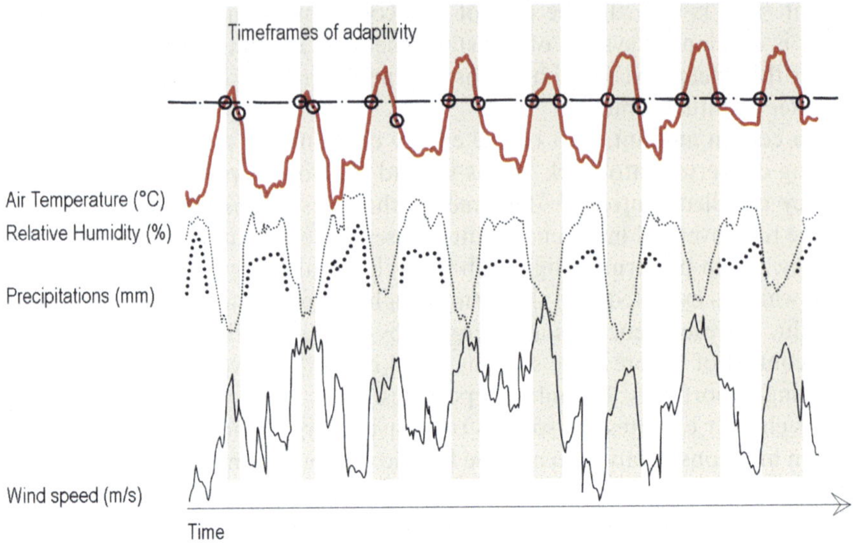

Fig. 4.1 Mapping of energy fluctuations and planned autoreactivity. This scheme hypotheses a way to proceed, in a design phase, to identify correlations between the energy sources and the reaction timeframes. Four different types of energy source fluctuations are shown over one week's time. The timeframes during which adaptivity is sought after are highlighted. In this specific example, the adaptivity timeframes correspond to the recurrence of temperature fluctuations: the intersections between the two form an almost horizontal dashed line. Hence, in this specific project, reaction to a thermal component is a promising solution to embed in the design

The forms of energy used by autoreactive systems can be found under two main forms: as contact and as action-at-a-distance forces. Contact forces involve physical contact between parts for force transmission. The category includes different kinds of frictional forces (friction, normal and fluid resistance). Action-at-a-distance forces, on the other hand, exert push and pull effects despite physical separation, and variate in intensity depending on the distance between the parts. These forces include gravitational and electric/magnetic forces.

Energy is typically subdivided into three main categories: radiant, potential and kinetic energy. The different manifestations of each energy typology are described within each category, and possible applications in the context of autoreactive systems are discussed.

4.2 Radiant Energy

Radiant energy is made of electromagnetic waves, which carries energy in its oscillating electric and magnetic fields. It is emitted in a wide range of frequencies, which can be visible or invisible to the human eye. When electromagnetic waves are absorbed by an object, the energy of the waves is converted to heat (thermal energy) or to electricity (in case of a photoelectric materials).

4.2.1 Electricity and Magnetism

The electromagnetic force is an action-at-a-distance force that is one of the fundamental forces we find in nature, together with the strong interaction (nuclear force), the weak interaction (radioactive decay) and gravitation. Radio and television waves, microwaves, IR-rays, visible light, UV light, X-rays and gamma rays are all different kinds of electromagnetic waves which differ in the frequency at which their electric and magnetic fields oscillate (Robinson and Kashy 2020).

The first field theories described electric and magnetic fields as separate, until these were found to be interrelated and to represent two different aspects of the same phenomenon. Electric and magnetic fields travel together through space, at the speed of light, as waves of electromagnetic radiation with the changing fields mutually sustaining each other, changing electric fields producing magnetic fields and vice versa, independently of any external interaction.

The electromagnetic force occurs between electrically charged particles. It is responsible for all phenomena occurring at a nuclear scale. In fact, all interactions between atoms can be explained with the electromagnetic force acting on the electrically charged nuclei and electrons, up to all forms of chemical phenomena (Helrich 2012).

Electrical current results from the transport of electrical charge within and between matters. The flow made up of charged particles (not molecules) produces an electrical current, and an electric field in the space surrounding it. Coulomb's laws states that *the magnitude of the electrostatic force between two charges is directly proportional to the product of the magnitudes of charges and inversely proportional to the square of the distance between them. The force acts along the straight line joining the charges. Like charges (++) repel each other while unlike charges attract each other (+−)* (Das 2013; Robinson and Kashy 2020).

While electric forces and fields act in the direction of the charge, magnetic forces and fields act perpendicularly to the field's source, influencing the charges.

Magnetism can (1) be found in some materials or (2) be created by electric currents:

- Ferromagnetic materials as iron, nickel, cobalt, few alloys, natural minerals (lodestone) are all minerals that exhibit magnetic proprieties resulting from the motion of electrons in the atoms of the material itself. These are strongly attracted to magnets and can be magnetized (some permanently), creating their own magnetic field.
- Electromagnets are made of a coil of wire (often wrapped around a ferromagnetic material to enhance the effect for more powerful fields), creating a magnetic field when electric current passes through it.

4.2.1.1 Ferromagnets

Magnets are common, low-cost, available in many different sizes and shapes, in different strengths, provide a force that does not wear out in time or with use, and always react in the same direction and magnitude. These proprieties allow magnets to have a wide range of applications in a design context, from toys and paper

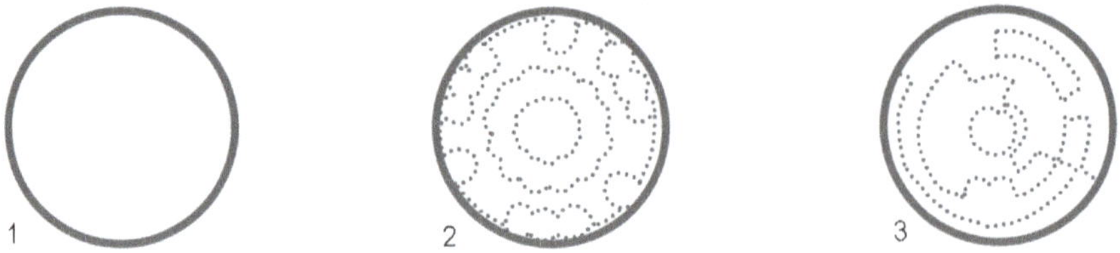

Fig. 4.2 Schematic representation of three different disc magnets with different design of the polarities (white represents negative charges and dashed lines positive charges): (1) standard magnet; (2) rotate-release magnet with alignment feature; (3) Hall-effect magnet producing a voltage difference used for sensors. Adapted from Correlated Magnetics (2019)

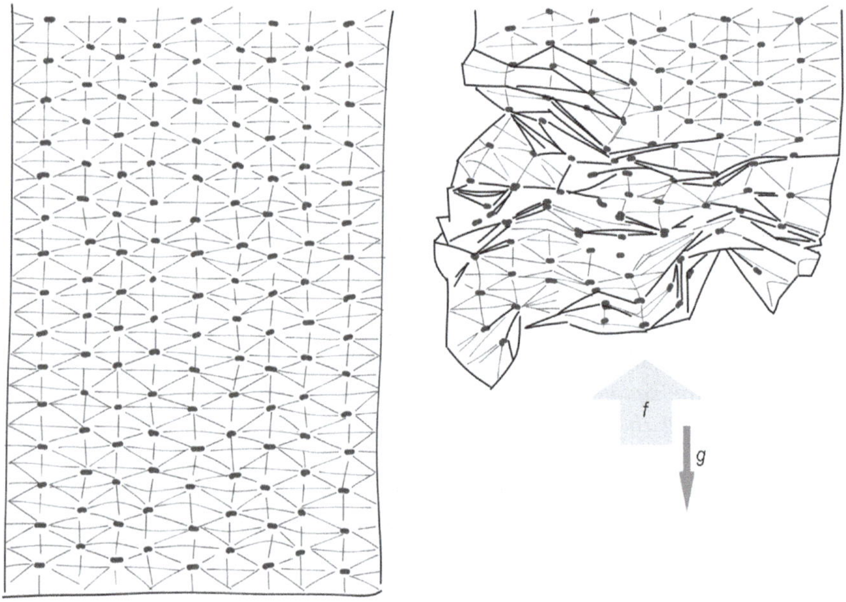

Fig. 4.3 Magnetic curtain represented in its undeformed state (left) and pulled up (right). In the second scheme, gravitational potential energy has been accumulated in the curtain as it has been pulled up through a direct force input (*f*). This potential energy is stored in the curtain as the magnets contrast the gravitational force (*g*) and thus the curtain's own weight

holders to components in electric generators and machines able to accelerate particles to speeds approaching that of light. Magnets are broadly used to enable temporary and totally reversible fixings.

Polymagnets are programmable magnets that, contrary to regular magnets which have a single north and south pole, have multiple north and south poles that can be tuned with varying magnetic fields and strengths to achieve a desired response and control different mechanical behaviours (Fig. 4.2).

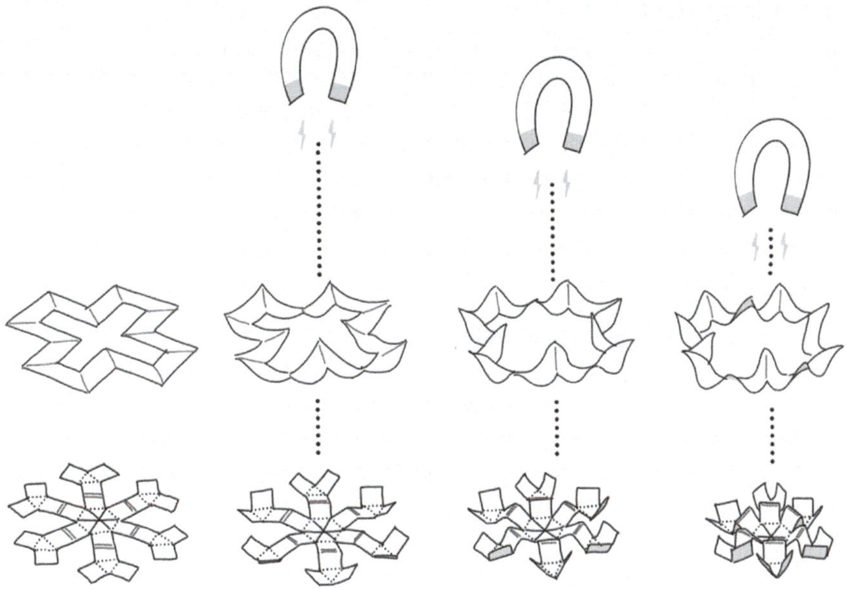

Fig. 4.4 Magnetic shape shifters developed at MIT, represented in four shape-shifting phases depending on the changing magnetic fields

From a design perspective, these smart magnets open up a wide range of new applications, incorporating an emotional response strongly characterizing the end product. Subtle differences in the feeling the product conveys can be tuned, from crisply snapping together to softly releasing, giving an augmented sense of control. Applications include magnetic hinges, attachments and couplings, dampening systems and positioning devices.

In the context of autoreactive systems, the magnetic proprieties of the materials can be used to intensify the effect of the device, or as an antagonist (opposing) force.

An example is the "Magnetic Curtain" (Kräutli 2008), made of a soft fabric subdivided into triangles by pre-folded lines integrating gold-plated block magnets. As the curtain is deformed by the user, according to the desired shape, the magnets attracted to each other allow the textile structure to retain its given shape (Fig. 4.3). The magnets act by storing the gravitational potential energy of the curtain to keep the set shape, counteracting the effect of gravity.

The magnetic proprieties of the materials can also be programmed to behave in certain ways, as is further deepened in ► Chap. 5. An example is magnetic shape shifters that change into pre-programmed shapes in response to an external magnetic field (Kim et al. 2018). By controlling the magnetic orientation of individual sections in the structure, these structures can be made to repetitively respond to fluctuations in the magnetic field shaping movements that mimic crawling, rolling and jumping (Fig. 4.4).

4.2.1.2 Magnetic Levitation

Magnetic levitation (or magnetic suspension) is a different use of the magnetic force than that previously displayed in ◘ Fig. 4.3 to counteract the effects of the gravitational (and any other) accelerations. The magnetic force provides a lifting force suspending the object in a specific configuration with no other support than the magnetic field. The object must then be stabilized to ensure that it does not slide or fall out of the area where the gravitational force is not neutralized.

In an autoreactive system, magnetic levitation could be used coupled with other means of autoreactive actuation to reduce friction forces.

4.2.1.3 Magnetic Self-assembly

Self-assembly refers to the spontaneous and reversible organization of units or components, that remain essentially unchanged throughout the process, into a pre-designed structure. The pattern of self-assembly is a direct consequence of specific interactions among the components, acting on a local level, without external direction. Once the assembly process is completed, and unless any other action is taken to separate the components, the formation rests in a state of equilibrium. In general terms, self-assembly avoids a one-at-a-time approach in the building of complex structures, simplifying piecing together parts difficult to access or too time-consuming to assemble.

Magnets are central tools in the design of self-assembling devices as they enable to achieve attraction, stabilization and alignment corrections in the positions of the moving components.

An example of this specific use of magnets comes from the M-Blocks self-assembling robots developed by researchers at MIT (Hardesty 2013). These small cubes with no appendages integrate a fast-spinning flywheel that flings the cube around as the flywheel is braked, transferring its angular momentum to the cube. This allows the cubes to perform different kinds of jumping, rolling and climbing movements in any direction, guided by specifically designed algorithms. Permanent magnets are the only feature that allow the cubes to align and attach to each other (◘ Fig. 4.5).

On each face of a cube, four pairs of symmetrical face magnets keep the cubes together and help positioning flipping cubes by snapping them into place as they land on the top of one another. The edges of the cubes are provided with cylindrical magnets that allow the cubes to pivot around the edge of one another. As the faces of the cubes touch, there is a slight gap between the edge magnets. This gap disappears as the cube pivots and the edge magnets touch, strengthening the pivot effect.

4.2.2 Solar Radiation—Light

Solar radiation is an electromagnetic radiation emitted by the Sun. The spectrum of different wavelengths is divided into three main regions: the ultraviolet (UV), the visible light—a very narrow part of the spectrum between 0.4 and 0.76 micron—and the infrared (IR)—over one-half of the energy emitted. Out of this,

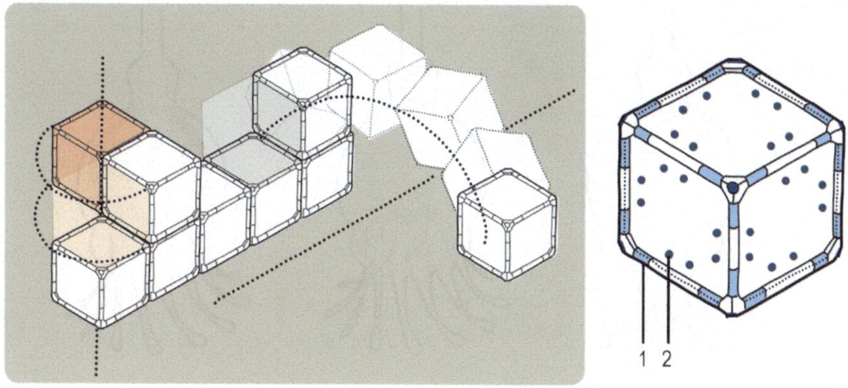

Fig. 4.5 Self-assembling M-Blocks. Pivoting and flinging motion of two cubes (left). Geometry of the cubes, with two different kinds of magnets (right): (1) bevelled edge magnets; (2) face magnets

the radiation that actually reaches the surface of the Earth is only a small part. Some wavelengths are selectively absorbed by Earth's atmosphere as most UV, IR and all wavelengths below 0.288 micron. Part is refracted by the clouds and water droplets hanging in the air and diffused back into space. The remaining part reaches the earth's surface in a diffused form, providing illumination even in the absence of direct sunlight. Diurnal and annual radiation patterns vary together with its intensity and duration depending on the varying thickness of air through which the rays must penetrate, due to the inclination of the Earth's axis to the plane of revolution (Zirin 2008).

The amount of radiation that falls on a given surface, in a specific place on Earth, is precisely predictable. For angles of incidence θ up to 45° with the normal (the line perpendicular to the surface at the point of incidence), the absorptivity of the surface is maximal. Under 45°, it drops progressively. Transparent bodies absorb a very small part of the radiation, reflecting and transmitting most of the spectrum. Opaque bodies absorb the radiation transforming it into heat (while a tiny part of it is reflected). Light and smooth surfaces generally absorb less radiation than dark and rough surfaces, which tend to heat up (Givoni 1976).

In an architectural context, the control of the reflection and/or absorption of radiation on vertical and horizontal surfaces is of great importance, as it has a major role in contributing to the thermal heat balance of the building (whether the effect is desired or not). The physical proprieties of the surfaces must, therefore, be given careful consideration, by designing surfaces with an appropriate angle of incidence, transparency, colours, textures, materials and orientation.

Different ways of using solar radiation involve (1) transforming the radiation into electricity through the use of photovoltaic cells; (2) transforming the radiation into kinetic energy with a radiation-reactive material as intermediary; (3) transforming the radiation into thermal energy. The first case is outside the scope of this review; the second case is further deepened in ▸ Chap. 5; the third and last case is further discussed in the following part.

Fig. 4.6 Thermobimetal lampshade. Lamp turned off (left) and on (centre). Section of the lamp showing the shape of the bimetal stipes anchored to the lampshade and the internal parts heated up by the waste heat radiating from the lightbulb (right)

4.2.3 Thermal Energy—Temperature Fluctuations

Thermal energy is the result of the random motion of molecules in a system and it is defined as the total of all kinetic energies within a given system. Temperature is the average kinetic energy within a given object and it is measured by three scales of measurement: Fahrenheit (F), Celsius (C) or Kelvin (K).

Heat is energy in transfer, as a thermal energy flow is caused by differences in temperature. If two bodies at different temperatures are brought together, heat will flow from the hotter to the colder body until equilibrium is reached. Heat transfers through radiation, conduction or convection.

4.2.3.1 Radiation, Conduction, Convection

Radiation is the transfer of heat through electromagnetic waves (mainly IR) irradiated by light sources as the sun, traditional light bulbs, irons, toasters, etc. It does not need physical contact to transfer energy.

An autoreactive example using radiative heat is heat-reacting lamps. These products integrate movement in order to add to the design a new dimension, conveying the feeling of the object being alive. While the primary function of the lamp is to emit light, the design of the lampshade resembles that of a flower that opens its petals as the waste heat from the light bulb, under the form of radiative heat, activates the bimetal strips in the lampshade (Fig. 4.6).

Conduction transfers thermal energy through the direct contact of solid particles. It takes place in all phases of a material, but works best in materials with simple molecules located close to each other, as in metals.

Convection is heat moving in a fluid (liquid or gas) as the fluid current moves from one place to another. It transfers the heat along with it.

An effect of Archimedes' principle (or law of buoyancy) is that hot fluids or gases rise relatively to cooler ones. What happens is that fluids/gases expand when heated, occupying a greater volume per unit mass and floating up just like an air

Fig. 4.7 The hanging mobiles, finely equilibrating each other, move with the smallest air movement (left). When the triggering factor within the surrounding environment ceases, the mobiles slowly returns to their original position. The fixings of the mobiles are a series of rods anchored to each other using small loops in the rods (right), allowing the rods to singularly move with two degrees of freedom, and the whole system to move with three degrees of freedom

bubble in water (Encyclopædia Britannica 2019a). Air in contact with the ground heated by electromagnetic radiation warms up through conduction and transfers to the upper layers through convection, creating turbulences as the currents move up (see next ▶ Sect. 4.4.1. Turbulence).

In architecture, thermal buoyancy is frequently used to cool down and ventilate the indoor spaces, in double skin facades, in chimneys and shafts. When openings are provided at different heights and the indoor temperature is higher than the outdoor temperature, the excess indoor pressure builds up at the upper opening where the air flows out, while a depression is formed at the lower opening where the air flows inwards. When the indoor temperature is lower, the flow direction is reversed (Oesterle et al. 2001).

The kinetic effects of thermal convection can be used in autoreaction to transfer the kinetic energy of rising air to objects making them move. This is, for instance, the case of Alexander Calder's hanging mobiles, kinetic sculptures powered by the air movements surrounding the models. These abstract sculptures, composed of a series of light metal plates equilibrating each other, assembled through metal sticks allowing them to move relatively to each other, hang from the ceiling, twisting and turning in thousands of variations according to imperceptible air movements (Fig. 4.7).

4.2.3.2 Phase Change

Heat is a thermal energy that is displaced from a thermodynamic system to another, caused by differences in temperature between the two. Heat flows from the hotter to the colder system until equilibrium is reached. For that to happen,

(1) one system cools down and the other heats up until they reach the same temperature, and in some cases, (2) one or both systems undergo a *phase change*.

Matter exists in three states or *phases*: solid, liquid and gas. What causes the bonds between molecules to break making a solid melt and a liquid to transit into a vaporous state or vice versa is the release or the imprisonment of thermal energy (see also next ▶ Sect. 4.3.4 on humidity and latent energy). When matter undergoes a change from one phase to another, thermal energy is added or removed from the system.

If an ice-cube is placed inside a drink, the drink—at a warmer temperature than the ice—will transfer its heat to the ice cube, and thereby cool down (which is the aim of putting the ice in the drink). The ice, on the other hand, will absorb the thermal energy of the drink until both reach the same temperature. The ice, therefore, undergoes a phase transition from solid to liquid by melting. Between the ice and the drink, no thermal energy has been lost, but the thermal energy of the liquid drink has been transformed and conserved in the water molecules through its phase change.

If on the other hand a cup of boiling water is placed next to a frozen swimming pool, the cup of water will transfer its heat to the swimming pool. However, although the cup of water is initially at a much higher temperature, the thermal energy (mass) of the frozen swimming pool exceeds that of the cup of water by many times and will end up freezing it. The water cup has changed phase by freezing, loosing thermal energy.

4.2.3.3 The Greenhouse Effect

The greenhouse effect generally refers to a system where light is transformed into thermal energy, which is then trapped in the system and heats it up as a consequence of not being allowed to dissipate. This effect can be found on a global scale, as the concentration of greenhouse gases in the Earth's atmosphere allows less radiation to be dispersed, re-irradiating it towards the Earth's surface (Hofman et al 2006). In buildings, the greenhouse effect can be either an unwanted condition or a sought-after strategy. In buildings in hot environments with inappropriate shading systems, the greenhouse effect can raise the interior temperatures dramatically above the comfort range. In buildings in cold countries on the other hand, it becomes a valuable free source of heat that can consistently contribute to the building's heat balance.

When sunlight falls on a building, the radiation that hits the opaque surfaces is mostly absorbed and transformed into heat. Heat on outside surfaces is then progressively transmitted to the interior of the building through conduction, with a time-lag depending on the thermal conduction of the building skin's materials. A greenhouse effect is obtained when the radiation shines through a transparent screen, as a closed window, and hits the indoor opaque surfaces heating them up. The glass, permeable to solar radiation but screening the thermal convection traps the heat inside the building, progressively heating it up.

Autoreactive systems activated by the greenhouse effects can mostly be found in greenhouses and in building facades, to ventilate rooms and shafts. This is the case of shading systems using paraffin-filled thermal cylinders that by expanding, opens double-glazed façade elements ventilating the façade (Auer and Molter 2019).

4.3 Potential Energy

Potential energy is the energy which is stored in an object because of its position in space. To gather this kind of energy in an object, an action is performed on it by an external force, working against the existing force field. As the force field tends to bring back the body into its initial position as soon as the external force is removed, potential energy is released. "Potential", therefore, refers to the capacity of the object to perform work as the stored energy is released (Cheremisinoff 2001). Traditional subdivisions are:

- Elastic potential energy;
- Gravitational potential energy;
- Chemical potential energy (as energy stored in fossil fuels or chemical reactions);
- Water vapour.

4.3.1 Elastic Potential Energy—Spring Force

The spring force is a contact force, which is stored in elastic materials as the result of their stretching or compressing. A classic example is the stretching of a spring. When a spring is not stretched or compressed, there is no elastic potential energy (or strain energy) stored in it, and it is said to be at its equilibrium position. From this resting position, the spring needs to be loaded with an input of kinetic energy (through compression or tension). The amount of potential energy stored is related to the material's proprieties and to the amount of the deformation: the more compressed/stretched, the stronger force is required to compress/stretch it further, and the more energy is stored. When the force acting on the spring is removed, the spring will return the accumulated potential energy in the opposite direction to which it has been loaded, restoring its initial position of equilibrium (Viegas 2005; Encyclopædia Britannica 2016a).

Similar systems able to store elastic potential energy are among others rubber bands, bungee cords, trampolines, springs that can be used in different mechanisms from catapults, arrow and bow, etc. The mechanical properties of an object determine how much of any strain energy is recovered in restitution as useful work. A variable used in many sports to estimate of the elasticity or the energy losses of an object relative to another object is the coefficient of restitution (COR). It is a dimensionless number calculated as the relative velocity of separation divided by the relative velocity of approach of two objects during a collision.

$$\mathrm{COR} = (\mathrm{bounce}/\mathrm{drop})^{1/2}$$

It is represented by a number ($1 > COR \geq 0$) with high coefficients of restitution representing elastic collisions with little wasted energy, and lower coefficients of restitution absorbing and dissipating most of the strain energy (Knudson 2007). For instance, a ball dropped from 1 m and rebounding 60 cm on a hard surface has a ($COR = 60/100)^{1/2} = 0.77$, the same ball dropped from 1 m and rebounding 40 cm on a soft surface has a ($COR = 40/100)^{1/2} = 0.63$.

In the context of autoreactive systems, elastic potential energy can become very useful as a side mechanism, loading up potential energy that is kept within the system for further use. Elastic potential energy can mainly be used in two ways:

- Using the spring force as antagonist in the movement cycle, moving the device back to its initial position once the force acting on the system is removed (see also ▶ Sect. 4.4. Kinetic energy);
- Combining elastic potential energy with kinetic energy, in a system reversibly converting kinetic to potential energy and vice versa, prolonging the duration of the movement (see also ▶ Sect. 4.4. Kinetic energy).

4.3.2 Gravitational Potential Energy

The force of gravity is an action-at-a-distance pulling force that all objects of mass have on each other, like the sun attracting its planets. Mass is the amount of matter that a living or a non-living thing has. On Earth, mass is equal to weight, as weight is a measurement of the gravitational attraction of a given object to the Earth (and Earth's gravitational attraction to the object). Therefore, if the same object was taken to the moon, its mass would stay the same, but the weight (now gravitational attraction to the moon) would change (Viegas 2005).

When two objects of mass are distanced, counteracting the gravitational pull they exert on each other (a ball is picked up from the ground), gravitational potential energy is gathered in the object, because of its position (height of the ball). So, gravitational potential energy (PE) depends on the object's mass (m), its height (h) and the acceleration of gravity (g). There is a direct relation between gravitational potential energy and the mass of an object (the more massive the object, the greater the potential energy), and the height of an object (the higher the distance of the object, the greater the potential energy). When gravitational force is released, the objects of mass are drawn to each other, producing kinetic energy (the ball is dropped down) (Encyclopædia Britannica 2016a).

In the context of autoreactive systems gravitational potential energy can be used:

- As antagonist in the movement cycle, moving the device back to its initial position once the force acting on the system is removed;
- Combined with kinetic energy, in a system converting kinetic to potential energy (see part on kinetic energy), to prolong the duration of the movement.

The latter is the case of playground swings and pendulum clocks that use the moving pendulum to drive the clock's mechanism. A pendulum is a suspended

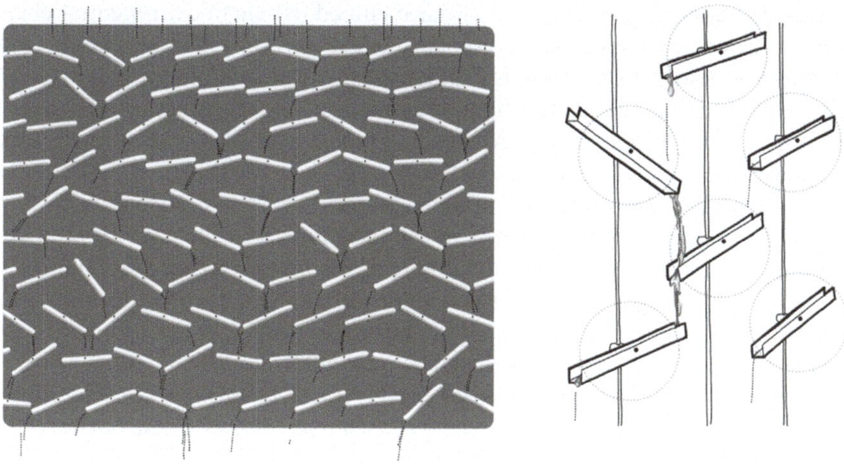

Fig. 4.8 The tipping wall system

object that is free to move between two extremes, with the central position between swings being the lowest point. As the bob (the weight) is lifted, it gathers potential gravitational energy, which is released when the bob swings down. At the bob's lowest point, all the gravitational potential energy has been transformed into kinetic energy which makes the bob swing in the other direction as it is released, shifting again to potential gravitational energy. The total energy of the pendulum is constant and changes back and forth between kinetic and potential energy. The motion would eventually continue endlessly, if no friction and air resistance made the swing stop (Viegas 2005).

The tipping wall, a monumental artwork created on a vertical outdoor surface (Kahn 2011), makes use of the gravitational potential of falling water to convey motion. The opaque wall surface is covered with metal tubes, hinged in the middle, and allowed to rotate with approximately 45° to the horizontal plane. Water runs over the facade, filling the tubes until they tip, empty and return to their original position (Fig. 4.8).

The metal tubes move in a slow pace, under the effect of their fluctuating centre of gravity as the tubes fill and empty. The system as a whole imparts the observer with a sense of a semi-chaotic rhythm, as a giant natural clockwork. The downside of this system, from an autoreactive point of view, is that the water that has reached the bottom of the artwork must be pumped up again to close the movement loop.

4.3.3 Chemical Potential Energy

Bonds between atoms store potential energy. Electrons do not provide any potential energy themselves, but the electronic charge differences that form the bonds exist as a source of potential energy.

Organic material, gas and coal all have embedded chemical potential energy. Much of the energy conveyed by biological organisms on Earth ultimately comes from the Sun, which releases energy under the form of light and heat waves. This energy is absorbed by plants through the process of photosynthesis and converted into food energy (Viegas 2005). When these plants are eaten, the animal's digestive system chemically breaks down the energy stored within it. Calories measure the amount of heat energy contained in food, so, when an organism burns calories, it is in fact using the energy in the food for its bodily activities (moving, breathing, heart beating, regulating body temperature, etc.) (Viegas 2005). In the same way, fossil fuels are nothing else than old sunlight trapped in millions of years old plants and organisms. When we burn fossil fuels, we are using the chemical potential energy contained in them (Benyus 2002).

The reason why chemical potential energy is not taken into consideration in an autoreactive system is that the energy exchange cycle is often not reversible and, most of all, the energy source is not renewable.

4.3.4 Humidity—Hygroscopy

Water vapour in the atmosphere regulates air temperature by absorbing thermal radiation from the Sun and from the Earth, gathering latent energy through the state change of the water molecules (see ▸ Sect. 4.2.3.2. Phase change). The latent energy in the atmosphere in form of water vapour is responsible for generating storms, precipitation (see ▸ Sect. 4.4.1. Atmospheric turbulence) and condensation (Encyclopædia Britannica 2013).

Absolute humidity is the density of the moisture content of the air. The capacity of the air to contain humidity is determined by temperature and depends therefore on geographic location and time. For instance, the air masses over the equatorial deserts contain vast quantities of water vapour, although relative humidity is very low because of the high temperatures (Encyclopædia Britannica 2013). Relative humidity (RH in %) is the ratio of the actual absolute humidity to the maximum moisture capacity of the air (absolute saturation) at a given temperature.

Hygroscopy is the ability of a substance to attract and hold water molecules from the surrounding environment, implying both kinetic and thermodynamic criteria of behaviour (Van Campen et al. 1980).

For instance, wood has important hygroscopic proprieties. The quantity of moisture contained in wood is subject to continuous change, in equilibrium with the ambient atmosphere. Loss of moisture content is called desorption, gain of moisture content is called adsorption. A secondary effect of the adsorption/desorption process is the consequent swelling/shrinking of the wood's volume, varying according to the direction of the fibres (Célino et al. 2014).

The structure of the wood's cell walls is organized with filamentous microfibrils oriented in the direction of the longitudinal axis. Moisture is absorbed by wood and as the cellulose molecules expand in proportion to the quantity of water adsorbed, the material swells as a result. This water is held inside the cells and after that the cell walls are saturated, and liquid water also enters cell cavities, allowing the moisture content in living trees to vary greatly from 30 to 300% (Garcìa Esteban et al. 2005).

Another natural material with high hygroscopic proprieties is wool fibres, which, for instance, are perceived significantly drier than less hygroscopic fibres at a comparable RH rate of the surrounding environment. The kinetic effects of these adsorption/desorption processes are, however, at the moment not significant enough for further analysis in this context (Plante et al. 1995; Downes and Mackay 1958).

In the case of autoreactive systems, it is possible to use the potential energy in terms of humidity by embedding designed motion induced in hygroscopic materials, sensible to humidity fluctuations. In this case, the swelling and deforming capacities of the material, undergoing repeated absorption and adsorption, are used to convey movement. This case will be further analysed in ▶ Chap. 5.

4.4 Kinetic Energy

Kinetic energy (KE) is the energy which an object possesses due to its motion. It is given by the mass (m) of the object and it is directly proportional to the square of the object's velocity (v). Having gained energy during acceleration, the body maintains the kinetic energy unless the speed changes due to another force's interaction, as in the case of friction.

Linear or translational kinetic energy can be calculated from the object's mass (m) and its velocity vector ($\mathbf{v}$). It is a scalar that describes how much work an object in motion can perform.

$$\mathrm{KE} = \frac{1}{2} m v^2$$

As can be seen in the formula, squaring velocity makes the energy of motion primarily dependent on the velocity of the object. This means that a light, but very fast-moving object, will tend to have more kinetic energy, and will create a higher energy collision if halted, than a little heavier but slow-moving object (Knudson 2007).

Newton's cradle is an example of how kinetic energy is transmitted, as it implies physical contact. Kinetic energy is imposed on a ball as it is swung through the air. As the ball collides with another ball, it stops suddenly and the ball it collided with accelerates as the kinetic energy is passed on to it. These collisions are of elastic nature, where kinetic energy is preserved. Had the collisions been inelastic, the kinetic energy would have dissipated in various forms of energy: heat, sound or breaking (Viegas 2005; Encyclopædia Britannica 2019b).

As seen previously mentioned in ▶ Sect. 4.3.2. Gravitational potential energy, kinetic and potential energy go hand in hand: together they make up what is called mechanical energy. The mechanical energy of an object is due to the motion of the object or to its position, eventually to both. Kinetic and potential energy have the propriety to reversely transform into one another, creating a

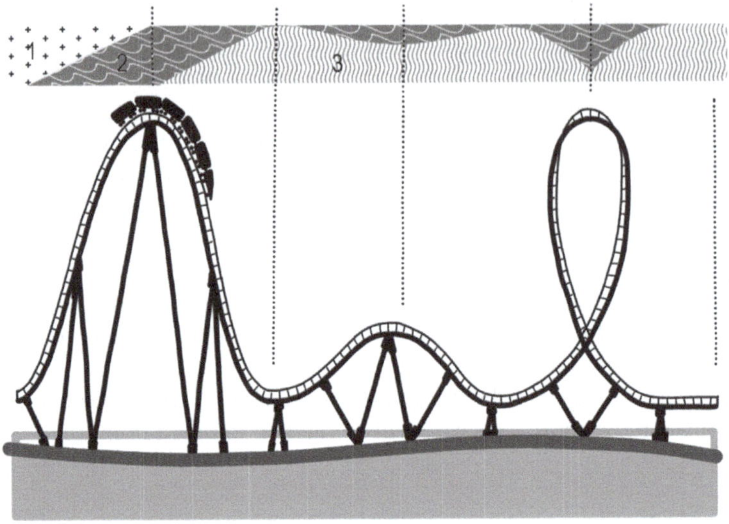

Fig. 4.9 Scheme of a rollercoaster with the coaster cars moving from left to right. The histogram over the scheme represents the different amounts and types of energy involved during the ride: (1) initial force input; (2) potential energy; (3) kinetic energy. Energy losses from friction are not included

continuous cycle of energy exchange. Many machines and objects use this process because the cycle requires (ideally) only an initial input of force, and then goes on switching from one state into another alternatively, as in a *yo-yo*.

Roller coasters mostly work without an engine, making a clever use of mechanical energy (Fig. 4.9). An external engine provides an initial force at the beginning of the ride, pulling the cars up a hill. Once at the top, the coaster cars possess gravitational potential energy which is released as the weight of the ride pulls the coaster downward. As the coaster speeds up, the potential energy is progressively transformed into kinetic energy. The rotational spirals along the path combine kinetic and potential energy, keeping the coaster in the mechanical energy loop, reversing kinetic and mechanical energy. The ride could go on forever if it were not for the friction forces, as the air friction and the friction of the coaster's wheels against the tracks. (Viegas 2005).

Unlike potential energy, there is only one form of kinetic energy. Different kinds of motion can, however, be defined:

- Vibrational kinetic energy, as when a drumhead or a metal pan is hit producing a vibrating sound and movement;
- Rotational kinetic energy, as when throwing a "lasso" (a loose rope) or spinning a spinning top;
- Translational kinetic energy found in linear movements from one place to another.

In an autoreactive context, especially outdoor environments provide many opportunities for unused kinetic energy to further convert and repurpose within a design context:

- Atmospheric turbulence, as in wind, air currents, draughts, etc.;
- Precipitations, as with rain, hail, snow, etc.;
- Direct interaction of living organisms, through direct physical contact.

4.4.1 Turbulence—Atmospheric Turbulence

In general terms, turbulent flows are characterized by chaotic motion in fluids, due to unorganized changes in pressure and flow velocity. Turbulence can be observed as unsteady vortices of many sizes appear and interact with each other, as in rivers, clouds, smoke, etc. While on the one hand, turbulence increases the energy needed for the fluid to flow through, it can be exploited to recreate complex movements that are difficult to predict.

Atmospheric turbulence under the form of wind can be described as moving air masses, flowing between two areas with different atmospheric pressure. On a global scale, the reason behind differences in pressure belts is the uneven incidence and distribution of the solar radiation over the Earth, which heats up some zones more than others. Air currents are created between high-pressure zones, as warm air expands and lifts, and low-pressure zones under the warm air, bringing in colder air.

There are three global belts of winds in each hemisphere: the trade winds, the westerlies and the polar winds. Other winds, as the monsoons, result from annual differences in heating of land and sea areas. Local winds occur over mountains and valleys; day and night breezes occur along shorelines (Givoni 1976).

Land and sea breezes are created as a consequence of water bodies heating up and cooling down more slowly than the earth. As in the morning land is heated by the sunlight before the sea, warmer air rises from the earth creating a negative pressure zone that allows the colder air above the sea to flow inland. In the evening, the process is reversed: the breeze from the sea stops around sunset and later at night the land breeze begins (Givoni 1976).

Mountain and valley winds are due to the temperature differences in the air over the sunlit slopes and the air over the shadowy valleys.

Adiabatic processes occur without transferring heat or mass between a thermodynamic system and its surroundings, but only as work (Bailyn 1994). Adiabatic cooling and heating in air masses happen due to changes in altitude that alter the temperature of the air (about 1 °C/100 m in altitude). Therefore, when a mass of air is pushed up a mountain, it moves from a higher to a lower pressure region and so expands and is cooled. Conversely, when an air mass descends, it is compressed and heated.

Dynamic inversion happens when two air masses meet, and the warm air is lifted up by the colder air. Surface inversion occurs during night as the air near the ground cools down faster than the air masses above it, creating stable conditions where no vertical air movements occur.

In urban contexts, buildings form an obstacle to laminar airstreams. Building facades with different orientations will, therefore, coexist under different pressure conditions: windward sides have pull (positive) pressures, and leeward sides are

4

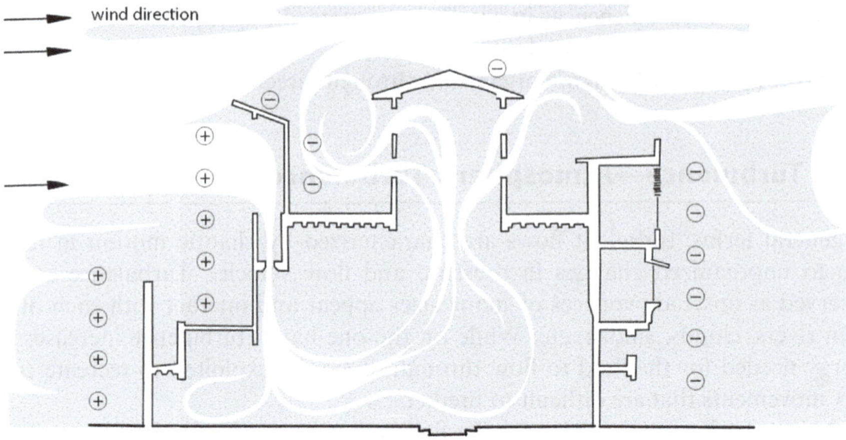

Fig. 4.10 Scheme of a section through the Qã'a of Muhib Ad-Dmin Ash-Shãf'i Al-Muwaqqi in Cairo showing positive and negative pressure surfaces and how the malqaf (Egyptian windcatcher) and the tower escape produce internal air movements. Adapted from El-Borombaly and Molina-Prieto (2015)

quite turbulent with push (negative) pressures. The roof is in all cases subject to negative pressure. The pressure on the windward surfaces is not evenly distributed, but it is stronger in the centre of the pressure zone and diminishes on the sides. When the angle of incidence of the wind on the exterior wall is around 45°, the negative pressure at the downwind corners almost disappears (Oesterle et al. 2001).

These differences in pressure are commonly used to passively ventilate buildings: when openings are provided in zones with different pressure, a stabilizing airflow is created between them, flowing through the building. The greater the pressure difference, the higher the potential force for ventilation. (Givoni 1976) (Fig. 4.10).

Living and non-living things moving in space also give rise to constant and sometimes imperceptible air movements. These air movements are of particular interest in indoor contexts, where turbulences occur in a more controlled way. This is the case of:

- Fans that aim to create a pressure difference between their front- and backside pushing the air forward while sucking more air into the fan;
- Doors closed quickly can create an instant turbulence;
- Elevators moving in their shafts can create whistling sounds while pressing the air during their ride.

In the context of autoreactive systems, kinetic energy in terms of wind flows and other turbulences can be used:

- To gather energy within a system, transforming kinetic energy into potential energy, winding up a mechanism which then releases (kinetic) energy with a specific purpose;

Fig. 4.11 Two sailing boats following a different course under the same wind conditions. The forward velocity reached by the craft (V_B) depends on the given true wind velocity (V_T) and the point of sail (boat course in relation to the wind). The apparent wind velocity (V_A) is the wind experienced by the boat in motion, and more specifically, it is the wind acting upon the leading edge of the most forward sail

- To convey the kinetic energy transforming one movement into a different organized movement with a specific purpose;
- To purely interact with the kinetic energy, with no other purpose than to make the movements visible to the eye.

The first case is that of the Strandbeest (BBC One 2010): engineered artificial animals made out of a system of plastic tubes and connections and animated by the wind transformed into kinetic energy (see ▶ Chap. 1). To initiate a first walking cycle, these mechanisms need to stand a few hours in a wind flow, building up enough pressure within the plastic bottles that are the motor of the mechanism.

The second case is that of sailing boats, that would use the linear motion of the wind flow to propel a craft on a chosen course (Fig. 4.11), which can be directed up to a 30° to the wind without losing propulsive force.

Another example in the same category is windmills. While the modern wind turbines transform kinetic energy into electricity, old windmills transform the linear motion of the wind flow into a rotational movement (Fig. 4.12) that would mill grain, pump water or both.

The design of the system in modern wind turbines has evolved by shaping the rotor blades as airplane wings, creating a lift that is turned into a rotating force. As the rotor shaft turns, it drives an electrical generator (Sclater 2011).

The third case perfectly describes the facade of the Pittsburgh Children Museum "articulated cloud" (Kahn 2004). It is an outdoor screen aimed at providing visual indoor comfort, composed by plastic tiles moving in the wind giving a beautiful aesthetic effect (see ▶ Chap. 1). In this case, the kinetic energy in

Fig. 4.12 Isometric drawing of the machinery of a Dutch hollow-post windmill with rotating four-bladed sails. As the wind struck the sails, a horizontal shaft geared to a vertical shaft was made to rotate, driving the machinery at the ground level. From Hoeft and Long (1976)

the wind is transferred into a movement of the tiles under the action of frictional forces simply for an aesthetic purpose.

4.4.2 Precipitations

Clouds are formed by water particles which are about one millionth smaller than a raindrop. Precipitations are provoked by all of these water particles, which cool down and bond in different shapes—drizzle, rain, snow, ice crystals, hail, etc.—and finally fall from the clouds down to the ground.

Rain is caused by adiabatic cooling of the cloud formations (Givoni 1976). Convectional precipitations are heavy showers of short duration that occur mainly during the hot seasons in tropical regions. Hot earth surfaces heat up humid air masses ascending them. When the air starts to condensate because of the altitude, latent heat is released, speeding up the ascent of the air masses even more.

Orographic precipitations come from air masses which are pushed over the slopes of mountains. Rainfall is mostly on the windward side and diminishes on the opposite side of the ridge.

Convergent precipitations occur when air masses converge at low pressures. These occur both in tropical regions with showery rains, and in middle latitudes with widespread and long rains (Encyclopædia Britannica 2016b).

Snow is formed by ice crystals which occurs from latitudes 35° N and 35° S to the respective poles and on higher altitudes, covering all in all about 23% of the surface of the Earth permanently or temporarily. Snowflakes generally have a hexagonal geometric pattern. Size and shape depend on the temperatures and on the amount of water vapour available. The more the air is humid, the faster the crystals grow, developing branches and clumping together one to another. The colder and drier the air, the more particles remain small and compact. Texture and density of fallen snow vary constantly, densifying or turning into ice which can cause snow slides or avalanches on steep hillsides or on surfaces when temperatures change (Encyclopædia Britannica 2016b).

Frozen precipitations occur in three types: graupel (granular snow or soft hail <0.5 cm), sleet (partly frozen ice pellets) and hail (hard spheres of ice of 0.5–15 cm). Hailstones are usually formed of alternating layers of ice, depending on the temperature of the areas it moves through.

In the context of autoreactive systems, precipitations can be used both through the kinetic energy component, as for wind flows:

- To gather energy within a system, transforming kinetic energy into potential (elastic) energy, winding up a mechanism which then releases (kinetic) energy with a specific purpose;
- To convey the kinetic energy transforming one movement into a different organized movement with a specific purpose;
- To purely interact with the kinetic energy, with no other purpose than to make the movements visible to the eye.

Moreover, the potential gravitational energy of the medium can be used by involving a fluid in its melted form (rain) and in some cases in a solid form (hail, snow), using the substance's mass to gather potential energy.

- This effect is similar to the first case, as energy under the form of potential gravitational energy is gathered within the system, winding up a mechanism which can then release (kinetic) energy with a specific purpose.

4.4.3 Direct Interaction—Applied Force

An applied force is a contact force where kinetic energy is applied onto an object by an animate or an inanimate system. Any action we take to move an object—ride a bike, throwing a ball, pushing a button, cracking an egg, etc.—makes that we act upon the object with an applied force.

In nature, the Venus flytrap (see part on motion schemes and patterns and case studies) is a carnivorous plant which reacts to the triggering effect of insects moving on its surface with a chain reaction closing the leaves and trapping the insects. Similar to this example is the case of the *Mimosa pudica*, which reacts to touch by closing its leaves to protect itself.

In traditional architecture, change and adaptation of the building mostly rely on the interaction of the users with their surroundings. Opening and closing of windows, doors, use of blinds and curtains rely on the kinetic effects applied on the building through human interaction.

Autoreactive systems are defined as systems which employ unused energy in the surrounding environment to move. Therefore, a system where the user provides direct energy input to produce work is excluded from the autoreactive classification. A bicycle, which requires direct interaction of the user to produce motion is not an autoreactive system.

Hence, kinetic energy deriving from direct interaction can be used in an autoreactive context only when:

- The applied force does not serve the autoreactive system primarily, interaction is not the primary purpose of the user;
- The applied force serves primarily the autoreactive system, but it is accidental (the interaction is unintentional).

An example of systems using interaction as energy source is kinetic floors that convert the kinetic and gravitational energy of people walking or jumping on them into electricity by integrating piezoelectric materials. Vertical movements of 5 mm become rotating movements driving a generator (Li and Strezov 2014; Safaei et al. 2019).

4.5 Conclusion

The indoor and outdoor environments surrounding us provide many opportunities of unused energy that can be employed to design autoreactivity. Within this context, this chapter has given an overview of the available unused energy sources available for designers. Different options of use have been suggested, with examples borrowed from the fields of art, design, architecture and engineering, and categorized following the standard classification of energy into radiant, potential and kinetic.

Out of the discussed energy sources, many, however, depend on factors beyond the control of the designer. While the human needs change at an hourly, diurnal and seasonal rate, the available energy sources in the surrounding environment change with a totally different frequency. Many of these sources, as atmospheric turbulence, humidity fluctuations and interaction are largely characterized by unpredictability and instability in terms of occurrence and magnitude. Potential conflicts, therefore, arise between availability of the energy source and the timing of the functional needs. Moreover, specific functions might not share any parameters with available energy sources, not allowing any reaction.

This is why understanding the underlying relationships and factors that are part of the synergy are of primary importance for designers to allow a conscious and clear approach to the autoreactive design. To achieve this, designers need to map the patterns of energetic availability flows and compare them to the design requirements to verify the compatibility in time and magnitude.

References

Auer T, Molter PL (2019) Robustheit und Autoreaktivität: Temperaturregulierung mittels Dehnstoffelementen. In: Schumacher M, Vogt MM, Cordón Krumme L-A (eds) New MOVE Architektur in Bewegung - Neue dynamische Komponenten und Bauteile. Birkhäuser Verlag GmbH 2019, pp 76–79. ISBN 978-3-0356-1353-7

Bailyn M (1994) A survey of thermodynamics. American Institute of Physics Press, New York, NY, p 21. ISBN 0-88318-797-3

BBC One (2010) Theo Jansen's Strandbeests. Wallace & Gromit's world of invention episode 1 preview—BBC One, Youtube. Retrieved 10 Feb 2015 from ► http://www.youtube.com/watch?v=H-SKyHmjyrkA

Benyus J (2002) Biomimicry: innovation inspired by nature. Harper Perennial, New York

Célino A, Fréour S, Jacquemin F, Casari P (2014) The hygroscopic behavior of plant fibers: a review. Front Chem 1:43. ► https://doi.org/10.3389/fchem.2013.00043 (eCollection 2013)

Cheremisinoff NP (2001) Potential energy. condensed encyclopedia of polymer engineering terms. Butterworth-Heinemann, pp 200–255. ► https://doi.org/10.1016/b978-0-08-050282-3.50021-4

Correlated Magnetics (2019) Polymagnetsm. Retrieved 1 Oct 2019 from ► http://www.polymagnet.com/polymagnets/

Das A (2013) Lectures on electromagnetism, 2nd edn. World Scientific Publishing Co. Pte. Ltd. ISBN 978-981-4508-26-1

Downes JG, Mackay BH (1958) Sorption kinetics of water vapor in wool fibers. J Polym Sci 28:45–67. ► https://doi.org/10.1002/pol.1958.1202811605

El-Borombaly H, Molina-Prieto L-F (2015) Adaptation of vernacular designs for contemporary sustainable architecture in middle east and neotropical region. Int J Comput Sci Inf Technol Res **3**(4):13–26. ISSN 2348-1196

Encyclopædia Britannica (eds) (2013) Humidity, Encyclopædia Britannica, Inc. Published June 17, 2013. Retrieved 19 Apr 2020 from ► https://www.britannica.com/science/humidity

Encyclopædia Britannica (eds) (2016a) Potential energy. Encyclopædia Britannica, Inc. Retrieved 30 Jan 2016 from ► https://www.britannica.com/science/potential-energy

Encyclopædia Britannica (eds) (2016b) Precipitation. Encyclopædia Britannica, Inc. Retrieved 2 Feb 2016 from ► https://www.britannica.com/science/precipitation

Encyclopædia Britannica (eds) (2019a) Archimedes' principle. Encyclopædia Britannica, Inc. Published June 26, 2019. Retrieved 19 Apr 2020 from ► https://www.britannica.com/science/Archimedes-principle

Encyclopædia Britannica (eds) (2019b) Kinetic energy. Encyclopædia Britannica, Inc. Published November 25, 2019. Retrieved 19 Apr 2020 from ► https://www.britannica.com/science/kinetic-energy

Galvàn I, Bonisoli-Alquati A, Jenkinson S, Ghanem G, Wakamatsu K, Mousseau TA, Møller AP (2014) Chronic exposure to low-dose radiation at Chernobyl favours adaptation to oxidative stress in birds. Funct Ecol 28:1387–1403. ► https://doi.org/10.1111/1365-2435.12283 Retrieved 26 Jan 2016 from: ► http://onlinelibrary.wiley.com/wol1/doi/10.1111/1365-2435.12283/abstract

Garcìa Esteban L, Gril J, de Palacios de Palacios P, Guindeo Casasus A (2005) Reduction of wood hygroscopicity and associated dimensional response by repeated humidity cycles. Ann Forest Sci 62(3):275–284. Retrieved 30 Jan 2016 from ► http://arxiv.org/pdf/physics/0503174.pdf

Givoni B (1976) Man, climate and architecture. Applied Science Publishers Ltd., London. ISBN 0-85334-108-7

Hardesty L (2013) Surprisingly simple scheme for self-assembling robots. MIT News. Retrieved 26 Sept 2019 from ► http://news.mit.edu/2013/simple-scheme-for-self-assembling-robots-1004

Harris J, Wiggington M (2002) Intelligent skins. Elsevier Ltd., Oxford. ISBN 978-0750648479

Helrich CS (2012) The classical theory of fields, electromagnetism. Springer, Berlin, Heidelberg. ISBN 978-3-642-23204-6

Hoeft KS, Long CG (1976) Beebe Windmill: isometric of machinery. Historic American Engineering Record, National Park Service; Record NY-67. Retrieved 01 Oct 2019 from ► https://commons.wikimedia.org/wiki/File:Beebe_Windmill_Isometric_of_Machinery_Long_Island_NY.jpg

Hofman DJ, Butler JH, Dlugokencky EJ, Elkins JW, Masarie K, Montzka SA, Tans P (2006) The role of carbon dioxide in climate forcing from 1979 to 2004: introduction of the annual greenhouse gas index. Tellus B 58B:614–619. ► https://doi.org/10.1111/j.1600-0889.2006.00201.xCi

Kahn N (2011) Tipping wall. Marina Bay Sands, Singapore. Retrieved 19 Apr 2020 from ► http://ned-kahn.com/portfolio/tipping-wall

Kahn N (2004) Articulated cloud. Pittsburgh Children's Museum, Pittsburgh, PA. Retrieved 19 Apr 2020 from ► http://nedkahn.com/portfolio/articulated-cloud

Kim Y, Yuk H, Zhao R, Chester SA, Zhao X (2018) Printing ferromagnetic domains for untethered fast-transforming soft materials. Nature 558:274–279

Knudson D (2007) Fundamentals of biomechanics. Springer, New York

Kräutli F (2008) Magnetic curtain. Retrieved 18 Sept 2019 from ► http://www.kraeutli.com/index.php/2008/01/31/magnetic-curtain/

Li X, Strezov V (2014) Modelling piezoelectric energy harvesting potential in an educational building. Energy Convers Manage 85:435–442. ► https://doi.org/10.1016/j.enconman.2014.05.096

Mattson MP (2007) Hormesis defined. Ageing Res Rev 7(1):1–7 (2008 January). Published online 2007 December 5. ► https://doi.org/10.1016/j.arr.2007.08.007, Retrieved 01 Feb 2016 from ► http://www.ncbi.nlm.nih.gov/pmc/articles/PMC2248601/

Plante AM, Holcombe BV, Stephens LG (1995) Fiber hygroscopicity and perceptions of dampness. Text Res J 65(5):293–298. ► https://doi.org/10.1177/004051759506500507

Oesterle E, Lieb R, Lutz MH, Heusler W (2001) Double-skin facades: integrated planning, building physics, construction, aerophysics, air-conditioning, economic viability. Prestel, Munich, London, New York. ISBN 9789111315906

Robinson FNH, Kashy E (2020) Electromagnetism. Encyclopædia Britannica, Inc. Published February 04, 2020. Retrieved 19 Apr 2020 from ► https://www.britannica.com/science/electromagnetism

Safaei M, Sodano HA, Anton SR (2019) A review of energy harvesting using piezoelectric materials: state-of-the-art a decade later (2008–2018). Smart Mater Struct 28(11). ► https://doi.org/10.1088/1361-665x/ab36e4 (IOP Publishing Ltd.)

Sclater N (2011) Mechanisms and mechanical devices sourcebook. McGraw-Hill. ISBN 978-0-07-170442-7

Van Campen L, Zografi G, Carstensen JT (1980) An approach to the evaluation of hygroscopicity for pharmaceutical solids. Int J Pharm 51–18. Retrieved 30 Jan 2016 from ► http://www.sciencedirect.com/science/article/pii/0378517380900459

Viegas J (2005) Kinetic and potential energy: understanding changes within physical systems. The Rosen Publishing Group Inc, New York. ISBN 1-4042-0333-8

Zielinski S (2014) Some birds adapt to Chernobyl's radiation. Science News. Retrieved 4 Mar 2015 from ► https://www.sciencenews.org/blog/wild-things/some-birds-adapt-chernobyl's-radiation

Zirin H (2008) Solar radiation, Encyclopædia Britannica, Inc. Published December 11, 2008. Retrieved 19 Apr 2020 from ► https://www.britannica.com/science/solar-radiation

Materials of Autoreaction

Contents

S. G. L. Persiani, *Design of Autoreaction*,
Design Science and Innovation, https://doi.org/10.1007/978-981-15-6178-8_5

Manipulation of the environment can arguably be considered as a natural trait of adaptation in a broad range of animals. Mankind remains however the undisputed leader in the field by the broadness of contexts and extent of the manipulation. The development of new materials and technologies is so important to the shaping of human civilizations that historians describe entire periods of time as defined by the predominant use of a material or a technology, from the Stone, Bronze and Iron Age to the Industrial and the Digital Age. The environments we are in contact with in our everyday lives are manipulated to such an extent that almost everything we come in contact with—from our clothes to our tools, from our buildings to the outdoor spaces, from the food we eat to the energy we produce—has been reshaped with purpose by the technological knowledge of the time.

In terms of material manufacturing and synthesizing however, mankind has up to today used essentially the same "heat, beat and treat" logics to develop every variation of material and product (Benyus 2002). Our industrial model has throughout the nineteenth and twentieth centuries been focusing on the effectiveness and automatization of supply chains to produce big (and rising) quantities of identical, ever-cheaper parts, with harsh consequences to our environment.

Nature on the other hand lives and operates in the same place and can therefore not afford to dislocate its factories. The manufacturing processes respect the balance of the environmental conditions, are life-friendly, occur at "room" temperature, avoid the waste of energy and resources, and work in a closed-loop system. Natural materials all respect a number of rules (◘ Table 5.1): these are self-assembled, temporal, completely degradable and recyclable.

Today, technology is on the brink of a new paradigm shift. The age of nanotechnology and gene manipulation looks with new eyes on the solutions Mother Nature has developed in billions of years of evolution and geological history, trying to import and copy solutions from a material to process level. Scientists become the architects of matter. Materials are designed with unique proprieties observed in natural materials, "hacked" and artificially designed for man-made applications. The results blur the borders between natural and artificial, designed and evolved, produced and grown.

5.1 The End of Mechanics

» There are many great scientists of a thousand years ago who would have had no difficulty understanding an automobile, an engine, a helicopter and certainly not the most advanced architectural system. The craftmanship would have been astonishing, but the principles straightforward with respect to an understanding of the novel material proprieties. If we were to show the same great scientist a television, a computer or a radar, it would have appeared magical to them. The difficulty for them would not have been one of complexity, but rather they would have been lacking the mental framework required to conceptualize such non mechanistic devices.
(Clarke 1984)

Table 5.1 Comparison of the basic design proprieties of different thermal actuators

Characteristic	Description	Example
Temporal	Materials are completely degradable, transformed out of matter taken from the surrounding environment, to which it goes back at the end of their life, to be integrated into the next process	Natural process of growth and decay
Multifunctional	More proprieties coexist in one single material; functional differentiation through use of surface forces	Blowflies' eggs are impermeable to water and permeable to gases
Ultralight	Natural materials are proportionally ultralight compared to their working load and resistance capacity to avoid unnecessary energy consumption and optimize the use of available material	Sandwich constructions, as bird skulls made of a foam-like substance between two enclosing membranes (Nachtigall 2005)
Self-mending	Material constructions are chemically stable but not closed systems	Vertebrate bones can regrow upon need
Additive formation in time	Materials are built slowly, in time, by adding substance to the same structure	Seashells place new material on the shell edge, and more on one side than the other, making the shell grow into a spiral (Picado 2010)
Compartment formation	Materials built as the juxtaposition of compartments allowing easy construction and mechanical resistance and safety: failure is contained within some compartments and does not risk the whole structure	Cell structures, muscular fibres, foam-like structure of vertebrate bones, etc.
Hierarchical	Built up as "case-in-case" structure	In tendons, collagen fibrils form a collagen fibre, which in turn form primary, then secondary and tertiary fibre bundles which make up the tendon (Kannus 2000)

Innovation in the fields of material sciences and engineering, developing technologies that react on more biochemical principles than mechanical ones where the required functionality is embedded in the material, is bringing us towards the end of the mechanical era and into a new one. A growing category of smart materials is being developed with propriety-changing features: materials that change one or more of their characters in reaction to influencing factors like light, heat, humidity, etc., imitate defensive adaptations in nature.

This opens up towards completely new concepts and proprieties in the things that we shape. As the new materials are engineered not on mechanical principles but on biomimetic ones, they become not only optimized and more sustainable, but also reactive and multifunctional. This on the one hand involves that functionality is not anymore achieved through addition of more parts with different functions, but it is embedded in the material itself (Menges and Hensel 2008): responsive materials become the artificial muscle fibres. On the other hand, new parameters and complexity are introduced. Specific performances and behaviours become time- and context-related as materials undergo successive transformations to which temporary capacities are connected.

It is within this context of rapid development of reliable and cost-efficient tuneable material proprieties that the integration of environmentally reactive proprieties is made increasingly possible in the design of everyday artefacts and tools, opening up the path for totally new autoreactive devices.

All materials respond to change in the environmental conditions: through dilatation, shrinking, ageing, changing phase, etc. These characteristics have in traditional engineering contexts always been considered as problems to avoid. Hence, the design of artefacts has been adapted to prevent damages and unpredictability due to these behaviours. For example, iron expanding and shrinking according to temperature has created problems to railways, and dilatation junctions are often used in most systems integrating parts into iron.

In multifunctional new materials however, these structural anomalies are essential. Breakdowns and instabilities are highlighted, amplified and developed to be used as a design advantage. In fact, high responsiveness is often found in arrangements near phase transition or critically close to a physical propriety's stability edge. The most common phase transitions are thermally driven instabilities in which rearrangements occur in the structure at a critical temperature (Tc), and least one physical quantity vanishes, appears or becomes discontinuous (Janocha 2007; Ritter 2007).

New materials with engineered proprieties are described in the state of the art with a broad variation of resembling terms, hinting at the end-function, the production process or the specific characters. These terms describe variations of entities within the same family of new engineered technologies with a common "blueprint concept", highlighting and focusing on some aspects more than others. To distinguish the materials employed in an autoreactive context from other functional materials, the definition of different material categories is necessary.

5.1.1 Functional, Nano, Smart, Adaptive—Engineered Materials

Materials with changing proprieties have been used by man since early history. Hot water poured over wood was used to induce it to swell and split rocks and quarrying stones. In the beginning of the Industrial Revolution, the first bimetals were used as thermoelectric switches (Ritter 2007). Present man-made high-tech materials are comparatively a very recent development in the field of material sciences, with energy- and resource-requiring manufacturing processes that need to be made more environmentally friendly (Menges and Hensel 2008).

"Functional materials" (FMs) is a broad categorization covering organic and inorganic materials, wherein several functions are interlinked at a molecular level and can perform specific functions when triggered by environmental changes. While a technological component is usually made of different parts and materials—each integrating a different function—will fail to perform its designed purpose if one part is taken out of the system or the component is disassembled, in functional materials the designed behaviour is intrinsic to the material, allowing them to preserve the same proprieties and capacities, even if the volume is subdivided.

- Passive functional materials have structural anomalies in at least one of their physical quantities, often associated with a structural phase transition within a finite range of environmental conditions, as in phase-change materials (PCMs).
- Active functional materials convert energy (thermal, electric, magnetic and mechanical) from one form to the other and respond to external stimuli without exhibiting physical anomalies. Examples are piezoelectric, magnetostrictive, electrostrictive materials and shape-memory alloys (SMAs).

"Nanomaterials" (nM) refer to materials of diverse matrices (e.g. polymers, glass and ceramics) with proprieties engineered on atomic, molecular and supramolecular level to obtain specific chemical, physical, electrical and optical features (Gruber 2011). Nanomaterials are mostly used for functional surfaces, as easy-to-clean "lotus effect" or photochromic coatings (Sauer 2010), and do not always integrate propriety-changing features. These materials are therefore not further deepened in the context of autoreactivity.

"Smart materials" (SMs) refer to materials within the category of functional materials (Brownell 2005) with one or more proprieties that can be altered. The terms "intelligent" and "smart" refer to material intelligence on a small scale but have however been broadly used and misused in different contexts referring to greater and more complex technologies, often incorrectly associating these materials with the field of computer sciences (Fox and Kemp 2009). SMs are used in unstable and dynamic environments to respond directly to external simulation by undergoing changes in one or more of their proprieties—chemical, electrical, magnetic, mechanical, thermal—in a predictable way and maintaining the reversibility of the proprieties in time. Some of these materials can change more than one propriety. Magnetorheologic fluids (MRFs) not only change their flow proprieties under the action of a magnetic field, but also change the electrical, thermal, acoustic and optical proprieties of their suspensions. Other smart materials make changes in both ways, as piezoelectric smart materials that generate electric charge when put under stress (compression or tension), and reversely perform work (changing shape) on application of an electric field (Ritter 2007).

"Adaptive materials" (AMs) are for the biggest part-specific SMs, but can also be found in other material categories, as in functional or natural materials. AMs respond to environmental change by undergoing a physical change (kinetic, optical, etc.) and achieve a new state of balance. AMs can be defined through three parameters characterizing their nature and kinetic behaviour (Persiani et al. 2016):

- Base material (polymers, alloys, wood, ceramic, etc.);
- Adaptive effect (bicomposite materials, shape-memory, shape-change, material absorbing, phase-change effect and electro-active);
- Geometry of the kinesis (linear or volumetric expansion, orientation change through translation/folding/bending or torsion).

5

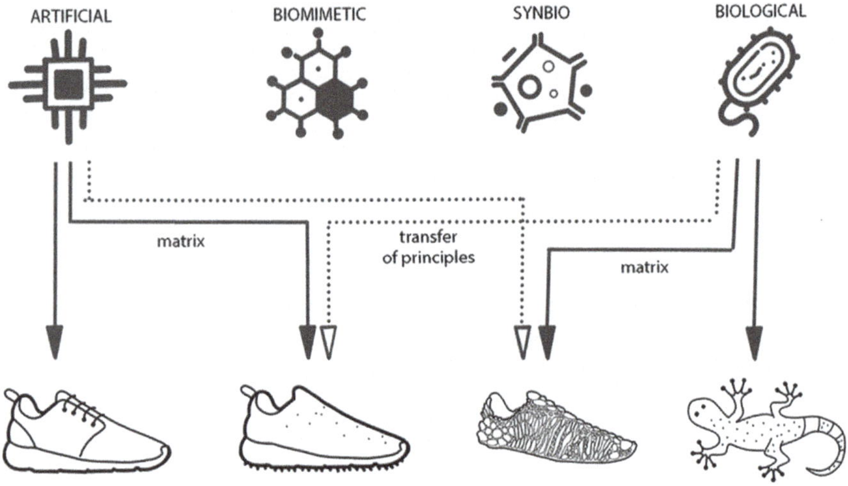

Fig. 5.1 Schematic representation of the development of materials following a biomimetic logic and a Synbio logic

5.1.2 Synbio—Biological Materials

Synthetic biology (Synbio) is a relatively new branch of study using biological engineering to manipulate life, programme organisms and synthetize artificial biological systems for a broad range of applications from engineering to medical applications. Synbio blurs the limits between living and non-living. Nature is no longer a simple inspirational model as in biomimetic material applications (like many nanomaterials), but becomes the object of fabrication, developing a new generation of genetically modified organisms (GMOs) that perform specific tasks with a predicted outcome (Persiani and Battisti 2018) (Fig. 5.1).

Using cutting-edge technology, Synbio enables the reconfiguration of living organisms, usually yeast or algae, to create man-made variants that are built from scratch, rewriting and programming synthetic DNA sequences. Instead of copying, cutting and pasting existing DNA sequences as previously done in genetic engineering, biological material is now broken up into hierarchically abstracted parts (basic building blocks) that can be modelled using sequencing and fabricating, with the support of computer-aided design (CAD) (Balmer and Martin 2008). Protocells are basic non-living molecules that can be chemically stimulated to respond to pressure, light, heat, but also to move, metabolize, reproduce, etc., just like any living cell.

The synthetic production of living material is for the moment still limited to basic applications but has the potential to revolutionize the way we make materials and tools. Not only could Synbio potentially revolutionize the complete life cycle of materials and products—growing of products with self-assembling, self-replicating, self-repairing, self-sustaining and self-degrading proprieties of living organisms—but also open up to new ways for us to envision our tools and spaces. Architecture and design could be embedded with completely new, previously impossible functions.

Synbio materials and systems are still in their very early stages of development; however, the possibility of transferring biological proprieties to artificial components forecasts a number of new functions among which:

- Reproduction. Controlling the reproductive timing and magnitude of artificial cells could revolutionize the whole life cycle of our products, from the way we manufacture, repair and re-purpose them. Materials could be grown on-site and could be autonomously healing and reacting to physical ageing and damage, simplifying also the replacement of broken and worn-out parts with new material (Persiani and Battisti 2018).
- Metabolic reaction. Networks of reactions could be engineered to release therapeutic compounds creating a new generation of monitoring substances inspired by the complex structures of mammalian cells (Collins et al. 2014).
- Mutation control. Biology is known to follow its own creative agenda. The evolution of Synbio artefacts on unpredicted paths is on the one hand the strength for these systems—as the selective advantage of a mutation is not foreseeable—and at the same time a source for big technical challenges and fears not to be able to contain "unwanted change". The use of microbial strains that are less susceptible to mutation (Collins et al. 2014) opens up towards bio-artefacts with controlled processes.
- Programmed cell death is the ultimate strategy to control cells not performing as engineered. In this case, the viral behaviour of the lambda phage bacterial virus is imitated, staying undetected until a program is activated to kill its host. More than a security measure, this capacity would allow the control of ageing processes on the one hand—developing artefacts that do not exhibit signs of age before their programmed end of use—and the programming of complex decay processes on the other (Persiani and Battisti 2018).

Using Synbio as a new design option, our artefacts would rapidly shift from mechanically dynamic systems to complex semi-living tools. In the present context however, Synbio systems can hardly be considered fully autoreactive. As any natural cell formations, also artificial cells require energy—in the form of chemical energy—to continue living and develop.

Rather than falling into the same classification, Synbio and autoreactive artefacts can be seen as parallel technologies that can be combined to enhance the features and add to the multifunctional complexity of one another.

5.1.3 Autoreactive Materials

Compared to the previously discussed material classifications, autoreactive materials (ArMs) fall into a quite broad family that ranges from natural (biological or mineral) to artificial (man-made or manipulated) origins. Autoreactive materials can be found in all the previously discussed categories; however, not all functional, smart, nano, adaptive or Synbio materials are autoreactive (■ Fig. 5.2).

5

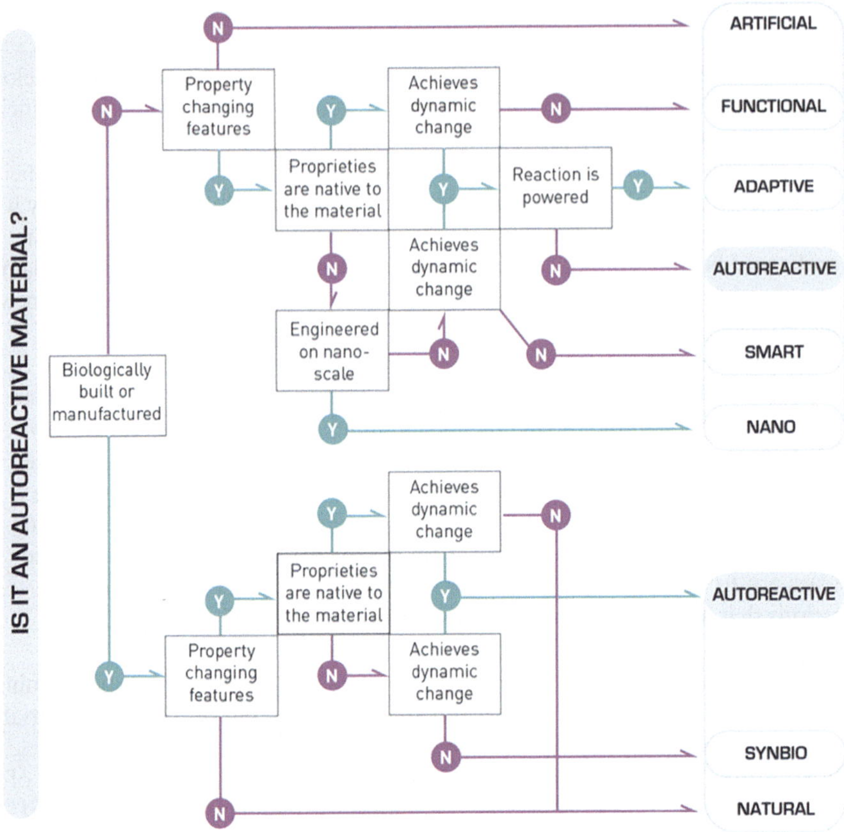

Fig. 5.2 Flow chart mapping autoreactive materials. If a material belongs to the autoreactive category, it is not excluded that it can also be part of other categories, as the functional, adaptive or Synbio categories. The opposite can however not be said, as all functional, adaptive or Synbio materials are not autoreactive

Autoreactive materials define themselves by a specific combination of input–output features:

- Input: the triggering action is a change in the surrounding environmental conditions, not a direct power input.
- Output: the reaction is a dynamic (kinetic) change in the material as a result of a shape or volume change.
- The movement is repeatable. The material can be returned to its initial state where the same triggering action will yield the same dynamic response.

The first condition excludes all materials (such as shape-change materials or piezoelectric materials) that react as a consequence of a voltage input.

The second condition excludes all functional materials that react to environmental change but do not achieve motion (as photochromic materials changing their optical proprieties or phase-change materials that change their thermal proprieties).

The third and last condition aims to exclude triggering actions that bring plastic deformations in the material as a consequence of structural damage. The conditions that are taken into account to define the autoreactive potential of the material must be within the operational range of the material so that all changes can be reversed, and the movement can be repeated.

When integrated into autoreactive systems, ArMs are usually used as mechanical actuators, selected and designed according to the physical principles they will respond to. Eventually, the combination of more ArMs also allows to design complex changes of behaviours to optimize energy and matter flows. Factors such as displacements and forces depend on the actuators' design, the scaling behaviour and eventual technical limits. Therefore, actuators must be chosen according to the dimensions of the device since the dimensions of the actuator establish which actuation principle is best suited for each specific application (Kohl 2004; Janocha 1992).

Existing smart and functional material classifications are typically systematized focusing on the obtained performance of the material (propriety change/energy exchange/matter exchange) (Sauer 2010; Ritter 2007). This appears logical as the complex performances of these materials are characterized by reactions to multiple inputs with varying outputs. ArMs on the other hand are characterized by multiple inputs and only one typical output (movement) and will therefore be further classified following the nature of the stimulation at the basis of the reaction. Hence, the materials explored in this chapter are classified and discussed within the three categories of energy sources previously defined in ▶ Chap. 4: radiant, potential and kinetic energy.

5.2 Radiant Energy

5.2.1 Electricity and Magnetism

A great number of smart and adaptive materials react to voltage with a kinetic output: electro-active polymers, magnetostrictive and shape-memory alloys, piezoelectric materials, etc. However, as previously defined, materials with a direct energetic input as voltage do not fall into the ArM classification and will therefore not be further explored.

As already discussed in ▶ Chap. 4, the magnetic proprieties of certain materials can be used to convey or implement motion in autoreactive systems (see also ▶ Sect. 5.2.3.4.4), performing mechanical movements without a motor.

5.2.1.1 Magnetic-Activated Shape-Change Materials (MA-SCMs)

Shape-change material (SCM) can assume a temporary shape when excited by a specific stimulus and subsequently recover their original shape as the stimulus is removed. While most SCMs are activated by a thermal stimulus, SCMs can also react to a broad range of other stimuli as light and humidity (see Sects. 5.2.2.2, 5.2.3.5 and 5.3.4.1).

Magnetic-activated SCPs are soft composite structures that consist of a silicone-based polymer matrix embedded with a magnetorheological fluid which introduces the shape-change proprieties. As the SCP is deformed and exposed to a magnetic field, it stiffens up to 30 times compared to its unstimulated conditions, retaining the given shape. As the magnetic field is removed, the material returns to its usual stiffness and the polymer bends back to its initial shape (Testa et al. 2019).

Applications for MA-SCPs are multiple, among which there are many heat-sensitive environments where temperature-activated SCPs cannot be used: medicine (stiffness-changing catheters), aerospace (self-inflating and folding vehicles), electronics (wearables) and robotics (performing mechanical movements without a motor) (Testa et al. 2019).

5.2.2 Solar Radiation—Light

Photosensitivity refers to a broad range of reactions that materials of different nature have when exposed to electromagnetic radiation. Photosensitivity can be found in organic materials—as in plant leaves that become darker or human skin that tans according to the intensity of the light conditions they are exposed to—and in inorganic materials alike—as in silver halides, commonly used in photographic film and paper.

Photosensitive or light-activated (LA) materials achieve a broad range of reactions, from optical change (photochromic materials), light emission (fluorescence and phosphorescence), voltage generation (solar cells), adhesion change (hydrophilic coatings combining light and water in a photocatalytic process), etc. Kinetic change is at the moment not a very common reaction in photosensitive materials. These are however under development: existing photosensitive ArMs are mostly found as shape-memory or shape-changing polymers.

5.2.2.1 Light-Activated Shape-Memory Polymers (LA-SMPs)

Shape-memory materials are functional, smart and adaptive materials that have been programmed to change and keep pre-defined shapes when triggered by specific stimuli. In the shape-memory effect, the material can be fixed into a predetermined shape and made to recover its original shape with different stimuli. The first generations of shape-memory polymers (SMPs) relied exclusively on temperature-activated solutions to fix the temporary shape (Kirillova and Ionov 2019). Light-activated shape-memory polymers (LA-SMPs) have since been developed with several advantages from using light to induce shape transitions: allowing remote activation, spatial control and recovery in appropriate wavelength.

LA-SMPs change the cross-linking density within the material at specific wavelengths (typically in the ultraviolet range) allowing the material to reversi-

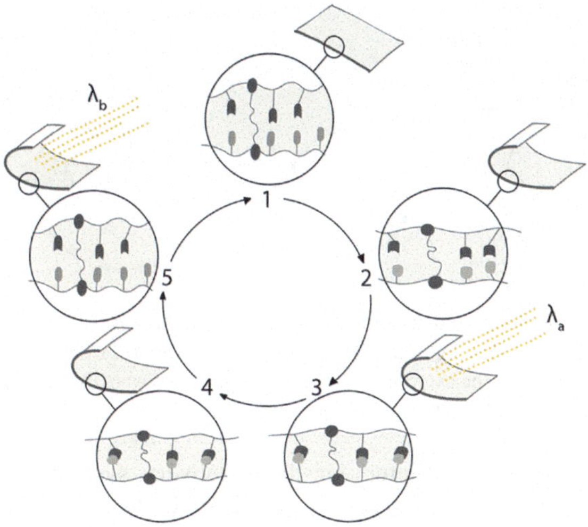

Fig. 5.3 Phases undergone by a light-activated shape-memory polymer: (1) material in its inactivated state; (2) stretching by mechanically deforming the material; (3) shape fixation by irradiation with the selected wavelength of light λ_a; (4) removal of the external stimuli, the material holds the same shape; and (5) photocleaving, releasing the fixation bonds by irradiation with the selected wavelength of light λ_b (adapted from Iqbal and Samiullah 2013)

bly switch between the mechanic proprieties of an elastomer and a rigid polymer (Havens et al. 2005). The samples' reaction range for temporary and recovery shapes is fixed by irritating the molecules with one wavelength of light, while a second wavelength of light reversibly cleaves and redefines the photo-cross-linked bonds (Lendlein 2010) (Fig. 5.3).

Applications of LA-SMP are found in the biomedical field (self-actuating prostheses), optical devices, flexible electronics, sensors, robotics (muscular replacements), aerospace (deployable and morphing structures) and smart textiles.

5.2.2.2 Light-Activated Shape-Changing Polymers (LA-SCPs)

Light-activated shape-change materials are closely related to LA-SMPs, in terms of materials used and advantages of light activation versus thermal activation. The big difference lies in the deformation, as the SCP sample deforms only when it is exposed to the stimulus, with its shape playing a major role in determining the geometry of the movement. As the stimulus is removed, the material recovers its initial shape (Fig. 5.4). The deformation can be repeated several times (Lendlein 2010).

Newer developments in material science open up to LA-SMP with improved proprieties as better shape recovery, biodegradability and biocompatibility (Xie et al. 2018).

As specific light spectrums can penetrate tissues without causing damage, the main applications for these materials are biomedical devices (self-expanding

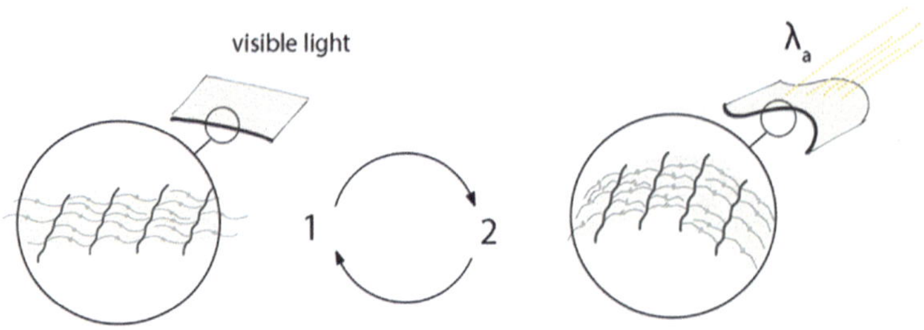

Fig. 5.4 Phases undergone by a light-activated shape-change polymer: (1) material in its inactivated state, irradiated with visible light; (2) shape change by irradiation with the selected wavelength of light λ_a; and (3) removal of the external stimuli, the material recovers the original shape as soon as the stimulus is turned off (adapted from Iqbal and Samiullah 2013)

stents, intelligent sutures, catheters or drug delivery systems). Moreover, NIR light can penetrate deep into tissues non-invasively without causing damage.

5.2.2.3 Light-Activated Liquid Crystalline Networks (LA-LCNs)

Liquid crystal (LC) refers to a state of matter with proprieties between the conventional solid and liquid, characterized for its self-organizing structural proprieties. Liquid crystals consist largely of organic molecules, although a few minerals are also known. Well-known examples of liquid crystals are soap solutions (Luzzati et al. 1957), biological cell membranes (Wojtowicz 1975) and DNA, which can form liquid crystal phases (Nakata et al. 2007). The protein mixture from which spiders generate silk is also in a liquid crystal state, as the ordering of molecules in the protein folding which is critical to its outstanding mechanical properties (Vollrath and Knight 2001).

LC materials may not always be in a liquid crystal phase but combine different proprieties and exhibit different phase transitions when exposed to different temperatures. Photo-responsive shape-changing liquid crystalline networks (LCNs) can finely tune their shape change in response to variations in the stimulus, as a preferential direction for deformation. Although the mechanical forces generated are quite low and the efficiency of photoenergy conversion is also not optimal, LCNs offer fast and large deformation (Iqbal and Samiullah 2013).

Researchers (Yu et al. 2003) developed a liquid crystalline elastomer-based (LCE) film able to selectively bend, moving parallelly to the irradiation direction of polarized UV light, combining the entropic elastic properties of elastomers with the self-organization of liquid crystals (Fig. 5.5).

Applications of light-responsive LC materials include a first light-driven autoreactive motor able to convert light energy directly into mechanical work by using a LCE-laminated film as a conveyor belt driving a pulley system (Yamada et al. 2008). The LC conveyor belt on one side is made to contract under exposure to UV light on one side, while irradiation with visible light on the other side of belt results in an expansion of the material. The combined effect of the expansion and the contraction movements, together with the mechanical belt–pulley system, results in a rotary movement of the motor.

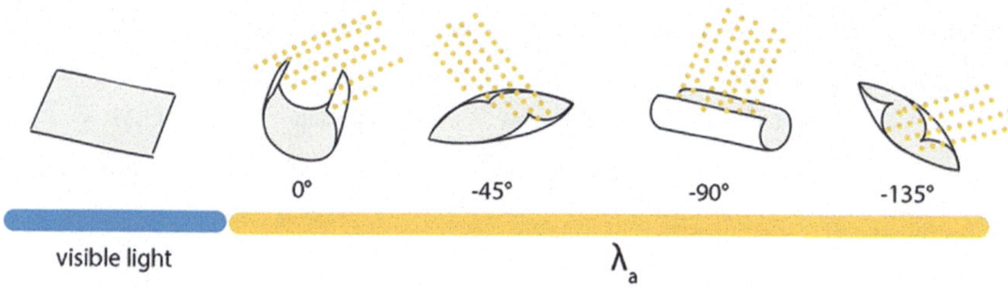

Fig. 5.5 Scheme of the direction-controlled bending of a liquid crystalline elastomer film with shape-change characters. The bending changes orientation, following the linearly polarized UV light at different angles, and goes back to the initial flat state as the light spectrum is changed to visible light (adapted from Yu et al. 2003)

5.2.3 Thermal Energy—Temperature

Temperature-activated (TA) kinetics is probably the oldest and most diffused type of autoreactivity that can be found. As shown by the first steam engines that convert the thermal expansion of water vapour into work, or the introduction of expansion joints in structures, the dynamic properties of materials in connection with specific temperature conditions have long been taken into account in various fields of engineering.

Two main categories of thermal actuation families can be distinguished: (i) materials with enhanced thermal expansion proprieties and (ii) materials with thermally actuated switches. To the first category belong thermal expansion materials (TEMs), thermo-bimetals (TBs) and thermo-bicomposite materials (TBMs). To the second category belong thermally activated shape-memory (SM) and shape-change (SC) materials.

Regardless of the type of actuation, thermo-activated materials are a quite established and affordable type of technology, and the material solutions are often already manufactured and sold in specific shapes (springs, films, wires, etc.) or as ready thermal actuator components (Table 5.2). Thermal actuators can efficiently replace sensor–actuator systems by integrating both functions within the material, transforming the thermal stimulus directly into mechanical work.

Applications for thermal actuators are commonly switches, ventilation systems, shading systems and regulatory and control mechanisms.

5.2.3.1 Positive Thermal Expansion Materials (PTEMs)

All matter in solid, liquid or gaseous form has the tendency to change its shape, area and volume in response to temperature variations. Therefore, temperature-dependent change can be found in organic and artificial materials alike and is described through the coefficient of thermal expansion.

Thermal expansion materials (TEMs) refer to materials with an important coefficient of thermal expansion, whether it is positive (PTEM) or negative (see ► Sect. 5.2.3.2), that allows them to undergo considerable changes in volume under specific ranges of temperature change.

Table 5.2 Comparison of the basic design proprieties of different thermal actuators (G-RAU 2015)

	SMA	TB strips	TB snap	TEM
Temperature–distance dependence	Abrupt	Continuous	Discontinuous	Linear
Max. operating temperature (°C)	80	40–550	350	≅110
Shape change	Elastic	Expansion	Expansion	Expansion
Movement	Traction, compression, torsion, bending, shrinkage	Bending and compression	Expansion and compression	Expansion
Work capacity	High ≅ 50 Nmm	Low ≅ 10 Nmm	Low ≅ 5 Nmm	Very high max. 15.000 Nmm
Reset	Counterforce	Automatic	Automatic	Counterforce
Work performed	Only when heated	Heating and cooling	Heating and cooling	Only when heated

5

Depending on the material, different material expansion proprieties can be used. In some materials, the change in volume can be proportional to the change in temperature, achieving a progressive change. In other materials, the volume might change suddenly as a specific temperature threshold is reached, obtaining an instantaneous and abrupt reaction.

The mechanics used to operate PTEM rely on the mechanics of single and multiple compartment systems, which take advantage of the volume change to do mechanical work. The materials are usually contained in rigid pressure-proof containers to control the expansion along a set direction (usually linear or rotational). The movement is reversed by the material itself as the temperatures decrease and the material shrinks, resetting the mechanism in its original position. In some cases, however, an antagonist mechanism (as a spring) is added to the mechanism to aid the recovering movement, which is generally less strong (G-RAU 2015).

Typical features which make of PTEM attractive materials to use in technological design are:

- Relatively cheap material costs (expansion waxes in particular);
- Possible long actuating path;
- High working loads;
- Not noisy;
- Sluggish reaction times (compared to other reactive elements).

PTEMs have a wide application range and can be found in many fields from architecture (windows, ventilation systems, sprinkler systems, thermostats,

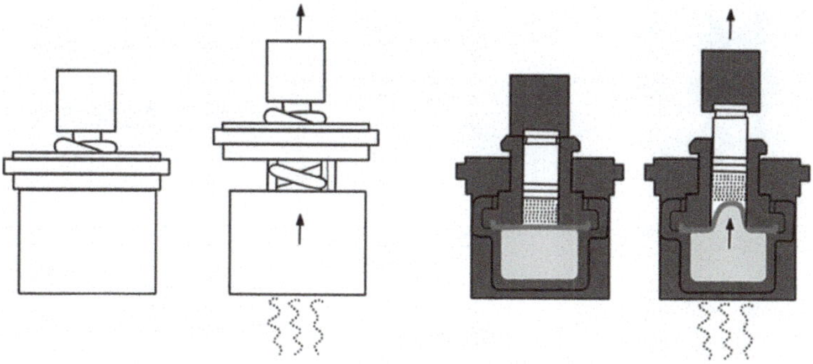

Fig. 5.6 Scheme of a thermostatic valve in its closed and open state, represented from the side (left) and in section (right). In the section, the PTEM actuator integrated into the valve is visible as it expands under the heated conditions, opening the valve

building technologies in general), agriculture (as in greenhouses) to automotive (operating valves for gases and liquids), etc. (Sauer 2010). They are commonly integrated into pressure-controlling devices, actuators, positioner drives or mechanical muscle elements as pistons (Fig. 5.6).

PTEM actuators exist in many sizes, with different expansion materials for different functions. More complex actions can be achieved through the assembling of more parts, using lever mechanisms, or by the incorporation of passive components of varying complexity (Ritter 2007).

5.2.3.2 Negative Thermal Expansion Materials (NTEMs)

Although rare, some materials contract upon heating due to the activation of different supramolecular structural mechanisms (Miller et al. 2009).

In the flexible network mechanism for instance, typical of the perovskite structure, dynamic deformation consumes open spaces in the crystal lattice yielding an overall net negative thermal expansion (Fig. 5.7).

NTEMs are mostly researched as compensators to neutralize and control the thermal expansion effect of other materials. By combining NTEM and PTEM, zero expansion composite materials or composite materials with controlled thermal expansion can be manufactured (Takenaka 2012). Therefore, these materials have high potential in structural engineering, dentistry (dental fillings), electronics (microchips), adhesives, fibre optics, aerospace technologies and high-precision optical mirrors (Miller et al. 2009).

In the context of autoreactive systems, the future development of tuneable thermal expansion of TEM composites could be of interest, as well as the possibility to endow a material with plural functions simultaneously.

5.2.3.3 Thermo-bicomposite Materials (TBMs)

Thermo-bicomposite materials consist of at least two permanently bonded layers of materials with different thermal expansion coefficients. The material with the lowest coefficient of thermal expansion is the passive component, and the

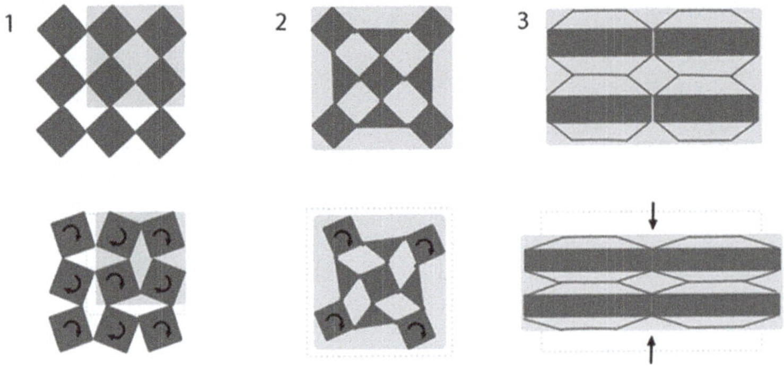

5

Fig. 5.7 Schematic representation of the negative thermal expansion mechanism in three materials: (1) perovskite structure shown as rotating squares; (2) zirconium tungstate structure shown as rotating squares; and (3) anisotropic expansion in silicates (adapted from Takenaka 2012; Miller et al. 2009)

material with the highest coefficient of thermal expansion is the active component. As the reacting temperature is reached for the active component, it expands while the passive component does not react, or reacts in a much smaller magnitude, working against the expansion. As the passive layer restricts the free movement of the active layer, the composite deforms, curving or bending. Any partial or complete restriction causes a corresponding force in the bending movement: the greater the expansion difference, the greater the effect (Ritter 2007).

The most common thermo-bicomposite materials are metals and alloys, which have been used since the Industrial Revolution. Newer technologies are however enabling the development of other variations using polymers, wood and mixed material solutions (as polymer and wood) (Ritter 2007).

Thermo-bimetals (TBs) are available as many different alloys. Common passive components are iron–nickel (invar) or nickel–cobalt–iron (superinvar). Active components are often manganese–nickel–copper (MgNiCu), iron–nickel–manganese (FeNiMg) or copper (Cu). Specific dimensional relationships must be maintained between the active and the passive components for the TBs to interact correctly. The combination of the different expansion coefficients and the various thicknesses can produce a wide range of grades in the bending movement.

The rate of change in shape is generally constant upon heating and cooling. Snapping effects can be achieved in certain limit conditions, and discontinuous behaviour can be designed by appropriately pre-bending the materials (G-RAU 2015).

Typical features which make TBs attractive materials to use in technological design are:

- Relatively inexpensive;
- Wide range of possible products;
- Predictable behaviour;
- High elasticity and strength, hot and cold ductility;
- Not noisy;
- Sluggish reaction times (compared to other reactive elements).

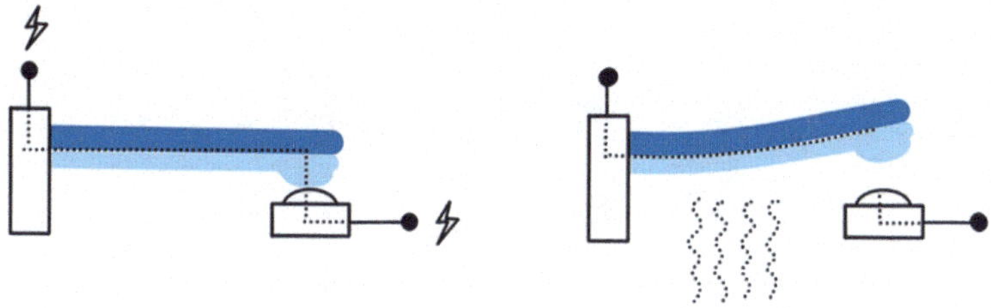

Fig. 5.8 Schematic representation of TB electric switch (left). As the active component reacts to a heat change, the TB curves interrupt the connection (right)

Having been on the market for many years, TBs are available in a wide range of forms from raw to intermediate or as end product. Most often however, TBs are available as laminate composites that come in bands or strips (normal strips, U-profile curved strips, reverse strips, spirals and helices) (Ritter 2007). When clamped at one or both ends, these can be used as actuator to create continuous, almost linear movement (Fig. 5.8).

TBs are also manufacturable in complicated shapes and assemblies for specific applications. They can be stamped in complex 2D shapes, that can be then folded to create 3D shapes, or also fabricated as special connections or 3D shapes (Ritter 2007).

Common applications for thermo-bimetals are thermostats, measurement and control systems, ventilation flaps, fire protection flaps and greenhouses. As actuation devices, TBs are integrated in creep action discs (continuous linear movement), snap-action discs (sudden linear movements) and can be used alone or in series to amplify the effect (Ritter 2007).

5.2.3.4 Temperature-Activated Shape-Memory Materials (TA-SMMs)

Contrary to conventional materials, where changes are due to elastic, plastic or thermal contributions, SMMs have other additional characteristics that are associated with the shape memory. As previously described in ▶ Sect. 5.2.2.1, the shape-memory effect allows the materials with this property to "remember" their original shape. After an apparent plastic deformation, SMMs are able to recover their undeformed shape in the presence of the right stimulus (Kohl 2004).

Thermally triggered SMMs are among the most common. Shape-memory materials may have different kinds of effect. The two most common are the one-way and two-way memory effects. In the one-way effect, the material can be deformed and fixed in a temporary shape, where it stays until it is heated above its transition temperature (T_{trans}) that activates the memory effect and recovery of the original geometry is observed (Lendlein et al. 2005) (Fig. 5.9a). In the two-way effect, the material remembers two different shapes, one at low temperatures and one at high temperatures (Fig. 5.9b). Technological advances now allow also the programming of multiple (usually triple or quadruple) shape memory.

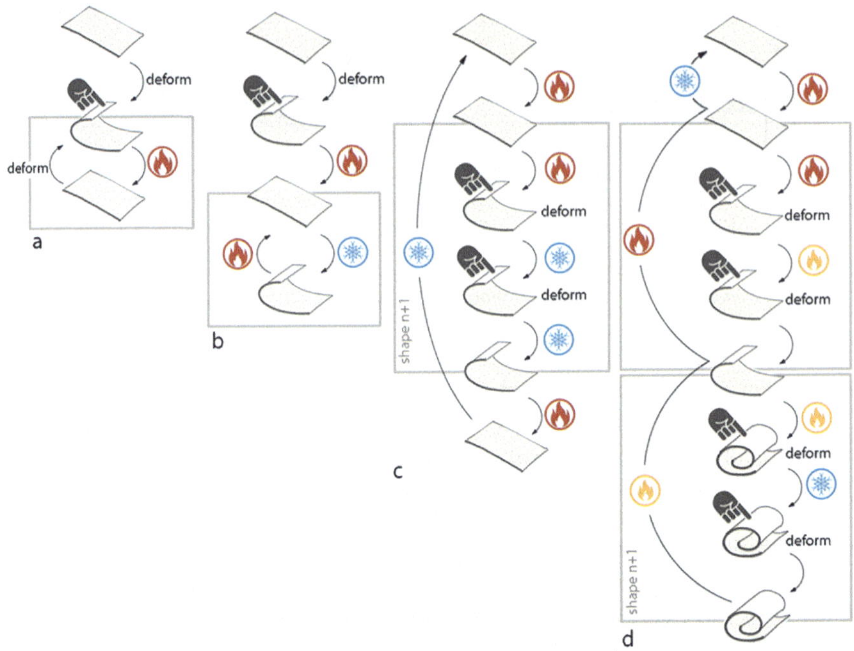

Fig. 5.9 Schematic succession of transformation phases in: **a** a one-way effect shape-memory cycle; **b** a two-way effect shape-memory cycle; **c** a dual-shape-memory cycle; and **d** a multiple shape-memory cycle with total shapes $(n+1)=3$ (adapted from Sessini et al. 2018; Zhang and Wang 2018)

The material is trained to remember an $(n+1)$ amount of shapes, whereby the material recovers all n shapes as the thermal conditions corresponding to each shape are restored (Sessini et al. 2018; Chen et al. 2015) (Fig. 5.9c, d).

Shape-memory proprieties are known in alloys, polymers, ceramics and biological systems, although the first two are the most common.

Shape-Memory Alloys (SMAs)

Shape-memory alloys (SMAs) are a small category of metal alloys mostly based on a nickel–titanium alloy, nitinol (NiTi), that can switch between crystallographic structures in response to a stimulus in the form of temperature or mechanical stress (Arun et al. 2018). After a thermomechanical treatment, SMAs take up the crystal structure (the shape) they were given earlier at a specific temperature: the alloy is said to be in an austenitic state at high temperature and martensitic at low temperature (Janocha 2007).

SMAs can perform a broad range of linear, rotational, bending, tension, compression, torsional, continuous/discontinuous kinetic effects depending on the component's geometry (G-RAU 2015). Moreover, SMA has super-elastic proprieties with 20 times higher elasticity at constant temperature than conventional materials (Ritter 2007).

Typical drawbacks are that SMAs are generally more expensive than TB alloys and that the working loads are generally quite low with the risk of showing fatigue effects. Therefore, SMAs are more efficient in small systems (Kohl 2004).

Features that make SMAs attractive materials to use in technological design are:

- Predictable behaviour;
- Simple design;
- Super-elastic proprieties;
- Not noisy;
- Long response time;
- Biocompatibility.

The SMAs that are the most interesting in an engineering and design context are the nickel–titanium (NiTi or nitinol), copper–zinc–aluminium (CuZnAl) and copper–aluminium–nickel (CuAlNi) (Ritter 2007; Janocha 2007).

Nitinol has a strong shape-memory effect, better tensile and elongation strength than other SMAs, and additional specific proprieties that can be achieved by adding alloys. It has been widely used and tested and has good market presence. CuZnAl alloys are relatively easier to machine than nitinol but have weaker shape-memory effect and poorer tensile strength.

SMAs are available as wires and rods (full section or hollow section suitable for carrying liquids which can affect the stimulation), springs, bands and strips (also usable as in multiple layers to multiply the effect), sheets (also usable to form surfaces where the shape recovery is however limited by the geometry) (Ritter 2007). SMAs can also be found in special shapes as clamps and stents (designed to fit specific requirements in complex 2D and 3D shapes), hooks and loop fasteners (integrated into stiff components and textiles as create reversible, complex, multidimensional connections) (Ritter 2007).

Applications for SMA are very broad, from vibration control elements (in submarines, ships and bridges), artificial muscles (robotics), electrical and automobile engineering, biomedical technologies (blood vessel widening technologies) and design (textile design and furnishings) (Sauer 2010).

Shape-Memory Ceramics (TA-SMC)

In shape-memory ceramics (SMC), shape-memory behaviour is exhibited in four different processes: viscoelastic, martensitic, ferroelectric and ferromagnetic (Arun et al. 2018). As this review focuses on materials that do not require the application of an electric field, only the first two processes will be further discussed.

Viscoelastic shape-memory ceramics introduce the shape-memory proprieties through the introduction of elastic strain energy into the rigid matrix of the material that is then released as a viscous plastic strain reversal upon reheating (Wei et al. 1998). These proprieties are found in mica-based glass ceramics with very good recovery of strain (Arun et al. 2018) that however strongly depend on pre-deformation temperature and rate, and on the reheating temperature and the holding time (Wei et al. 1998).

Martensitic phase transformation between two crystallographic phases is induced thermally (the shape-memory effect) or by the application of stress (super-elasticity) (Lai et al. 2013), resulting in transformation plasticity or transformation toughening (Wei et al. 1998). Martensitic transformations in ceramics have long had issues with mismatch stresses and shape distortions in adjacent crystal grains triggering fractures in the material (Lai et al. 2013). Examples of ceramic materials showing these proprieties are zirconia-based or partially stabilized zirconia (Lai et al. 2013; Kohl 2004).

Shape-memory ceramics however occupy a different region in property space than any other shape-memory materials, with a number of unique features that make SMCs very attractive materials to use in technological contexts (Lai et al. 2013):

- High strength allowing operation at very high stress levels;
- Dissipation of exceptionally large amounts of energy upon fracture;
- High refractory potential allowing operation at high temperatures.

Areas of application for temperature-activated SMCs are hydraulics, energy damping systems and high-temperature applications (Lai et al. 2013; Wei et al. 1998).

Shape-Memory Polymers (TA-SMP)

Shape-memory polymers (SMPs) represent a very dynamic field of study within material sciences, with a great variety of existing SMPs and a multitude emerging rapidly (Huang et al. 2010). The mechanism behind the shape recovery in polymers is very different from the phase transitions in alloys and ceramics. It relies on the viscosity and the interactions between monomers that affect the mobility of polymeric chain as it changes from a strained configuration to a less complex configuration (Arun et al. 2018).

SMPs have numerous advantages that make them attractive alternatives to SMAs, from the better possibilities to tailor the material proprieties with a broad range of application temperatures, to the lower material and manufacturing costs (Lendlein et al. 2005). Among the other features that make SMPs attractive materials to use in technological design (Huang et al. 2010; Lendlein et al. 2005) are:

- Lighter than SMAs;
- Higher elastic deformation than SMAs;
- Higher recoverable strain than SMAs;
- Wider shape recovery temperature range than SMAs;
- Possible triggering by multiple stimuli (even simultaneously);
- Broad range of products available;
- Possible to programme multiple shape memory;
- Biocompatibility;
- Biodegradability.

While pure SMPs exhibit lower mechanical proprieties (strength and stiffness) than metals and ceramics, selected properties or performances can be upgraded by combining the polymers with one or more materials/nanofibres/microfibres/particles to form shape-memory polymer composites (SMPCs) (Arun et al. 2018).

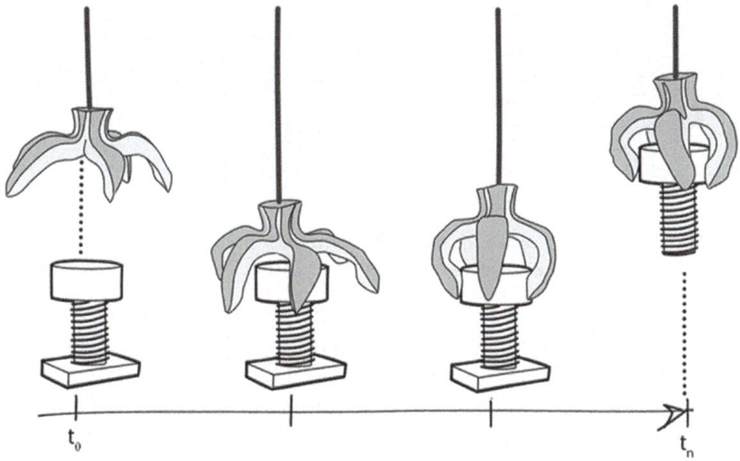

Fig. 5.10 Gripping sequence of a 4D-printed temperature-actuated SMP (adapted from Ge et al. 2016)

Other advancements in SMP technology include temperature-sensitive bio-based solutions (bio-SMP), developed as biodegradable and biocompatible SMP alternatives, typically synthesized from natural oil-based polyols (Zhang et al. 2015). Bio-SMPs allow adjustable transition temperatures, high degree of elastic elongations and acceptable mechanical strength (Petrović et al. 2017).

On an experimental level, SMPs have also been used as 3D printing material, achieving three-dimensional products able to change their features in time (four-dimensionally) (Han et al. 2018; Ge et al. 2016) (Fig. 5.10).

Due to the growing importance of polymeric materials in daily life, the range of applications is extremely broad, from biomedicine (minimally invasive surgery, implants, prostheses, bone repair, exoskeletons) (Lendlein 2010, Lendlein et al. 2005), microelectromechanical systems (Wan 2010), aerospace and automotive engineering (engines' choke system, morphing wing systems, self-deploying sun sails), high-performance textile design (wrinkle-free clothing and adjustable clothes), self-repairing plastic components and active disassembly mechanisms (He et al. 2013).

SMP foams are one specific class of smart polymers that combine the physical advantages of foams (lightness, porosity, 3D structure) with the advantages of shape memory to enhance their compressibility and recovery force to adapt to changing dimensional needs (Santo 2016). SMP foams are therefore mainly used in aerospace and biomedical devices, but also in microelectromechanical systems and actuators.

Shape-Memory Gels (TA-SMG)

Temperature-sensitive hydrogels are soft wet materials with properties similar to human tissue that mainly consists of water (Xiao et al. 2015). Most hydrogels morph uniformly, shrinking when heated and swelling when cooled, by absorbing

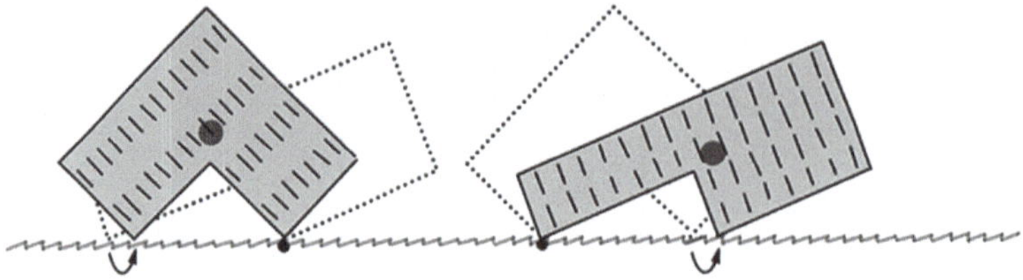

Fig. 5.11 Walking scheme through the thermally activated deformation of an L-shaped symmetric hydrogel with a layered structure (adapted from Kim et al. 2015)

or releasing water (see ► Sect. 5.3.4.2) (Han et al. 2018), and therefore mostly need to be operated within liquid solutions. The polymer switches between a hydrophobic and a hydrophilic state depending on whether the temperature is above or below the reaction threshold.

For example, poly(N-isopropylacrylamide) (PNIPAAm) is a well-known thermo-sensitive polymer gel which exhibits discontinuous volume change (Maeda et al. 2011). It undergoes a reversible temperature-induced phase transition from a swollen hydrated state to a shrunken dehydrated state, losing about 90% of its volume.

Advances in material engineering have however allowed the development of shape-shifting temperature-sensitive hydrogels under a magnetic field, operable when surrounded by air. The hydrogels selectively expand and contract without altering the water absorption resulting in a much faster shape change that allows the material to behave like an artificial muscle, mimicking, swimming and crawling self-propulsion. The gel is formed by a layered structure of parallel electrostatically charged nanosheets repelling each other and thus resisting orthogonal compression but deforming along parallel shear (Kim et al. 2015). As the temperature changes, the repelling force becomes stronger or weaker, changing the distance between the nanosheets rapidly, lengthening and shortening the gel (Kim et al. 2015) (Fig. 5.11).

In general terms, applications of temperature-controlled SMGs are potentially broad ranging from biomedical devices (drug delivery, biosensors, scaffolds for tissue engineering), to robotics (soft robots, flexible sensors, actuators), to different fields of engineering (microelectromechanical/nanoelectromechanical systems, microfluidic devices, self-healing coatings, temperature-controlling valves) (Arun et al. 2018; Xiao et al. 2015).

Shape-Memory Hybrids (TA-SMH)

Shape-memory hybrids (SMHs) represent integrated systems of multiple materials that combine the proprieties of individual materials to meet intermediate property requirements. With the hybridizing process, shape-memory materials in the form of thin films, fibres, wires, particle fillers or bulk matrices are integrated or joined with other materials to produce new composite materials with unique features able to enhance or actively tune their functionality or static and dynamic properties in response to environmental change (Arun et al. 2018).

SMHs are, as SMMs, based on a dual-domain system. SMHs often combine a material which is always in its elastic domain, with a material that undergoes a transition changing its stiffness depending on the stimulus. Therefore, provided that the properties of the selected materials are known for the respective domains of reaction, the design and fabrication of the composite are quite straightforward, allowing users with limited scientific/engineering background to manufacture many SMHs on a do-it-yourself (DIY) basis (Huang et al. 2010).

The SMMs were initially integrated (as the addition or the reinforcement) within monolithic or composite host materials (the matrix), but can also be used as the matrix, depending on the sought-after effect (Wei et al. 1998). The composites are either embedded within one another or bonded together in order for them to work as one. Chemical interaction between the materials in any of their phases tends to be avoided or minimized to facilitate the prediction of the properties and behaviours of the hybrids (Arun et al. 2018).

5.2.3.5 Temperature-Activated Shape-Changing Carbon Fibre (SCCF)

Shape-changing or shape-shifting materials are closely related to SMMs. The difference lies in the deformation, as the SCM samples deform only when exposed to the stimulus, with its shape playing a major role in determining the geometry of the movement. As the stimulus is removed, the material recovers its initial shape and the deformation can be repeated anew (Lendlein 2010).

The physical properties of carbon fibre composites are strongly influenced by the matrixes that hold the fibres together, from rigid or soft sheets to flexible spring effects. When combined with materials with shape-changing properties, the high stiffness, tensile strength and low weight characteristics of the carbon fibres are added to the stimuli-changing properties of the SCM.

By using 3D printing methods, researchers are able to target selected parts of the material's microstructure, programming it into performing specific dynamic effects. By selectively applying SCM according to specific geometries on the carbon composite, the material is made to twist or bend according to the pattern produced by the printer (MIT Self-assembly Lab 2019).

Due to the structural proprieties of the carbon fibre, shape-changing carbon fibres (SCCFs) are mainly driven by aircraft and automotive industries to replace hydraulic actuators, motors and hinges.

5.3 Potential Energy

5.3.1 Elastic Potential Energy

Elastic potential energy is rarely the source of energy provoking a reaction in an adaptive material. Generally, elastic potential energy is however involved in the building up of the energy within the given system (material or mechanical), which is then released upon a specific triggering agent.

In the context of material systems, the elastic potential energy is mostly involved in the autoreaction in two possible ways, which most of the time work together:
- As energy amplifier, elastic energy is built up through many small successive force inputs, which are then released in one single, concentrated action, as when drawing a catapult.
- As antagonist force to return the material into its initial shape or position.

In fact, many autoreactive materials, such as shape-changing alloys and polymers, exhibit in some phases very good elastic proprieties that are fundamental to the environmentally triggered reaction.

5.3.2 Gravitational Potential Energy

Gravitational potential energy is possessed by a material, as for an object, because of its position in space in relation to the gravitational forces. As for the elastic potential energy, gravity cannot directly trigger a change in the material. On the other hand, a propriety change of the material can be used to build up gravitational potential energy for further use. By changing shape and mass, or shifting its gravitational centre, a material can effectively use gravitational potential energy as a counterforce to amplify its movement. Gravitational energy can be used to:
- Building up potential energy in the material by counteracting the gravitational force;
- As antagonist force to return the material into its initial shape or position.

5.3.3 Chemical Potential Energy

As previously discussed in ▶ Chap. 4 (see Sect. 4.3.3), up to today we have extensively used the chemical potential energy contained in matter and in fossil fuels, by burning them.

A completely new class of synthetic and bio-based materials able to translate chemical recognition into macroscopic movements is developed by material scientists: Chemo-responsive or chemo-mechanical materials (CRMs). These materials react, mostly with a volume expansion or constriction, to a variety of chemical stimuli from specific pH values, to ionic strength of a solution and to biological cues (as antigens, to specific organic compounds, to enzymes, glucose, etc.) (Darvishmanesh et al. 2015).

5.3.3.1 pH-Responsive Polymers

pH values, that measure how acid or basic a fluid solution is, are often used as chemical switches being involved as indicators in a great number of biological (health-related and developmental) and environmental processes (environmental acidification and contamination).

pH-responsive (anionic or cationic pH-sensitive) materials respond to the changes in the pH of the surrounding medium by swelling, collapsing or changing other proprieties as when undergoing liquid-to-gel or soft-to-stiff gel transitions. The shape change is achieved in these systems as the changing pH adds or removes charges in the solution, creating charge repulsions and a change in the hydrophilicity of the material (Maerk et al. 2015).

pH-reactive shape-change polymers can also be synthesized by doping selected flexible biopolymers with organic compounds. A chitosan-based biocompatible polymeric film able to reversibly bend by applying high and low pH solutions on the same material was developed by researchers. Chitosan, a biocompatible macromolecule abundant in nature, known for its swelling and shape-memory tuneable properties, was sequentially deposited on the polymer, enabling specific animation capabilities through the distribution of the material (Kan et al. 2017).

Applications of the pH-reactive chitosan polymer, which is still in its initial phases of development, have been explored as edible dynamic fibres able to augment the sensory experience and flavour retention.

5.3.3.2 Biomolecule-Responsive Hydrogels

Hydrogels are cross-linked three-dimensional polymers with hydrophilic proprieties that allow these materials to absorb significant amounts of water, achieving dynamic change. Biomolecule-responsive hydrogels undergo structural changes in volume in response to the concentration of a specific biomolecule, as saccharides and proteins that characterize several physiological changes (Miyata 2015).

Due to their large water content, hydrogels are currently the synthetic materials that most resemble living tissue and are therefore extensively researched in the fields of biomedicine (tissue engineering, drug delivery, surface coatings, contact lenses, wound dressings, etc.).

5.3.3.3 Chemical Robots

Most autoreactive materials perform single shape changes upon external stimulation. “Chemical robot” gel actuators have been developed that can spontaneously achieve autonomous living-like motion without external driving stimuli (Maeda et al. 2011). The energy driving the repetitive motion comes from a cyclic nonlinear oscillating chemical reaction, a Belousov–Zhabotinsky (BZ) reaction, which acts like an artificial metabolic process generating chemical waves within the material (Maeda et al. 2007). The dissipation of the chemical energy results in an oscillatory kinetic motion forming spatial patterns through a reaction–diffusion system. Three different case studies were developed, focusing respectively on an inching locomotion, a ciliary locomotion and a peristaltic movement.

In the first case, an ionic polymer gel is synthetized by combining PNIPAAm (the matrix) and ruthenium monomer (the catalyst of the BZ reaction) with each component being distributed as a gradient in the polymer network of the gel. This type of distribution of the components allows to regulate the swelling rate of the gel to perform the wished motion. Additionally, an aqueous solution is prepared, containing in its substrate the reactant to the catalyst component. When immersed in the solution, as the substrate penetrates into the polymer network, the BZ reaction occurs in the gel propagating along its length as a chemical wave, producing a sequential localized bending and stretching motion before the next wave appears (Maeda et al. 2011). The substrate on which the gel lies is provided with an asymmetrical ratchet surface to prevent the gel from sliding backwards, resulting in a forward inching motion. The chemical wave diffusion scheme was also used to design an oscillatory ciliary motion and a periodical peristaltic motion by enhancing the local swelling regions propagated in the gel (Maeda et al. 2011).

5.3.3.4 Shape-Memory Biological Systems (SM-Bio)

As previously described in ▶ Sect. 5.1.2, systems integrating biological material to evolve new materials mostly follow two different logics (Persiani and Battisti 2018):

(i) The biomimetic approach, where synthetic materials imitate biological functions and integrate biocompatible substances;
(ii) The Synbio approach, where biological material is engineered to create new functions by hybridizing natural and synthetic materials and processes.

The first case is the one of SMPs triggered by biological activity. These SMPs are able to react by changing shape, under constant temperature conditions as is required in many cell cultures, when exposed to selected enzymatic conditions. The material is a polymeric composite created by blending fibres of two different polymer matrixes: the shape-fixing component poly(ε-caprolactone) (PCL), which is sensible to a specific enzyme, and a polyether-based polyurethane thermoplastic (Pellethane), which is a strong enzymatically stable elastomer (Buffington et al. 2019). As the material is subjected to mechanical stress, it takes on a temporary deformed shape due to the shape-fixing proprieties of the PCL component that counteract the elastic proprieties of the Pellethane (◘ Fig. 5.12a). When exposed to the specific enzymatic cocktail, the PCL crystallites are degraded allowing the Pellethane to recover to its programmed shape (Buffington et al. 2019) (◘ Fig. 5.12b).

For the application in an autoreactive context, enzyme-responsive materials (ERMs) still need to be further developed. Theoretically, the ERM can achieve a shape change able to produce direct mechanical force; however, the recovery process is at the moment too slow (a week under the highest enzymatic concentrations to recover to its programmed shape) (Buffington et al. 2019). Moreover, the movement is not repeatable once the enzyme-reactive component has been degraded. Therefore, applications for these materials are at the moment limited to biological and medical applications (drug delivery, tissue regeneration, stem cell culture, biosensors).

A very similar case, which however engineers a biological substance to achieve the same goal, is that of DNA hydrogels. These water-filled polymers are held together by strands of DNA and combined with a selected enzyme that has the ability to sever specific DNA strands when activated by a specific stimulus

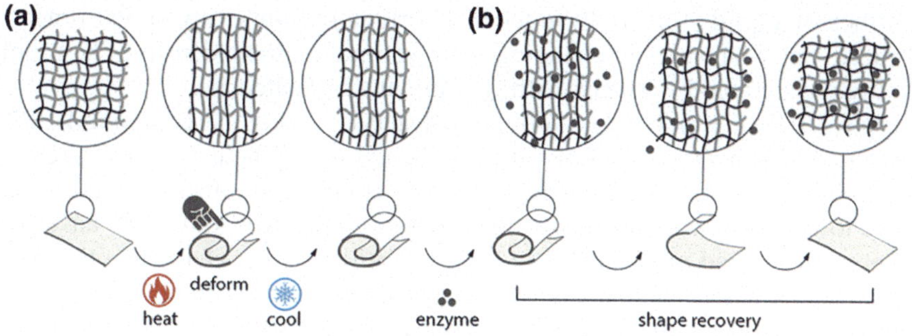

Fig. 5.12 Scheme of the shape recovery of an enzymatically triggered polymer: **a** shape-fixing phase and **b** shape recovery phase (adapted from Buffington et al. 2019)

sequence (Callaway 2019). As a result, when properly stimulated, the enzyme is activated snipping predetermined DNA strands, triggering a shape change or the dissolution of the hydrogel.

As in the previous case of ERM polymers, the shape-memory sequence in the DNA hydrogels does not seem to be reversible. Therefore, also these materials are so far developed for their capacity to recognize set genetic material in biological laboratory samples (Callaway 2019).

5.3.4 Humidity–Hygroscopy

Humidity-driven and hygroscopic materials use the swelling of the material structure upon intake of water to vehicle kinetic energy. A great number of structures in nature rely on these mechanisms, particularly plants that adapt to changing environmental conditions and activate seed dispersal when the conditions are most favourable.

These types of mechanisms have generally long reaction times as energy is built up over longer periods of time. Therefore, the potential hygroscopic energy can be used in two main ways (Persiani 2018):

- Action and reaction movement can follow the humidity fluctuations in the environment, achieving kinetic change on potentially long timeframes. In this case, motion is mainly achieved by a reiteration of the hot and dry environmental cycles achieving an intermittent change between the material's geometrical configurations.
- Potential energy can be stored within the material if coupled to an elastic or a spring mechanism. The swelling of the hygroscopic material draws a mechanism accumulating elastic potential energy, which can be released all in one action, snapping back as the balance is broken.

5.3.4.1 Absorbent and Super-Absorbent Polymers (AP/SAP)

Absorbent polymers (APs) are synthetic hydrophilic 3D cross-linked polymers that can, in short time, absorb and retain (even under pressure) important quantities of liquids (up to 30 times their original volume, up to 500 times for deionized water), reversibly changing their volume or their density. These polymers are regenerated by contact with still, moving or heated air.

Typical APs are silica gels, sodium polyacrylate (organic compound), PNIPAAm (see also Sects. 5.2.3.4.4 and 5.3.4.2), nylon 6. These polymers are most often found as composite granules, which are usually contained in bags, but can also be found as powder, bands or films (Ritter 2007). APs and SAPs have good market presence and are available in large quantities. Other characters that make them attractive materials to use in technological design are (Ritter 2007):

- Operation over a large temperature range (−10 to +80 °C);
- Material-dependent absorption capacity and speed;
- Maintenance-free;
- No volatile toxic elements.

APs and SAPs are primarily used as water-tightening systems or to prevent leakage, and not for their shape-changing proprieties. Common fields of use are agriculture (technical membranes and green roofs), hygiene products and in systems preventing water leakages (as in underwater high-voltage cable installations) (Ritter 2007).

Some absorbent polymeric materials, specifically hydrogels, are researched with the more specific aim of developing the kinetic properties.

5.3.4.2 Shape-Change Hydrogels

As previously discussed (see Sects. 5.2.3.4.4, 5.3.3.1 and 3.3.2), hydrogels strongly rely on liquid and humid environments to operate as they tend to morph uniformly in all directions by absorbing or releasing water.

PNIPAAm is a polymer with known hygroscopic proprieties that is often used as matrix for further material engineering. This hydrogel exhibits a large and reversible volume change in water around a transition temperature (T_t) of 32–35 °C. Under this temperature, the material has a hydrophilic behaviour with an extended coil structure, which leads to water uptake and swelling; above this temperature, the coil structure becomes very compact, resulting in a hydrophobic behaviour and significant shrinking (Han et al. 2018).

Improved macroscopic kinetic control of these materials can be achieved by relying on the microstructural design of the monolithic structure rather than on the chemical proprieties of the constituent materials. Directional swelling can be embedded within the material's microstructure by locally programming the orientation of stiff reinforcing elements throughout the monolithic structure. The combined action of the swelling action of the matrix and the directional orientation of the reinforcements can create a multitude of bending and/or twisting configurations.

To achieve this effect on a microstructural scale, researchers embedded successive layers of UHMR alumina platelets within a PNIPAAm structure by aligning them with the help of a rotating magnet. During the drying and water loss of the hydrogel, the rigid geometry of the reinforcement restrains the freedom of movement of the material, resulting in a tunable bending movement. Drying brings the hydrogel back to its original geometry (Erb et al. 2013) (▫ Fig. 5.13a).

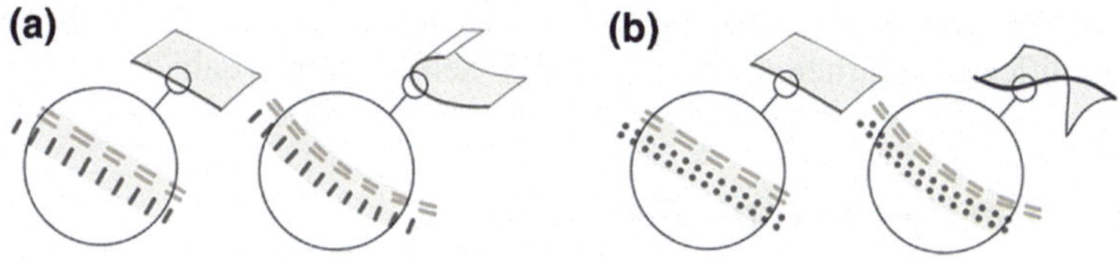

Fig. 5.13 Scheme of two different combinations of the orientation in the reinforcement of the hydrogel, resulting in: **a** bending and **b** twisting (adapted from Erb et al. 2013)

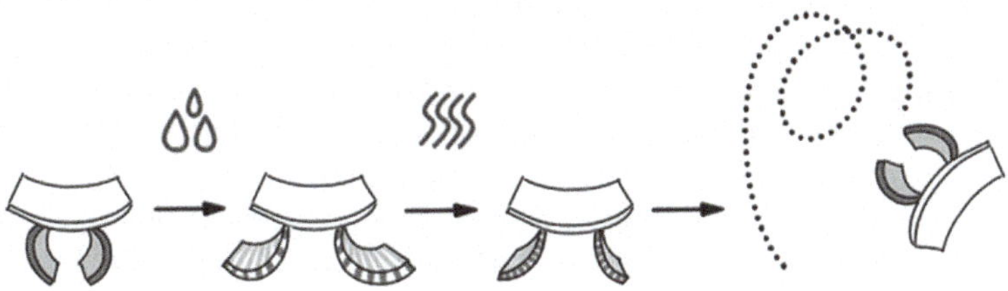

Fig. 5.14 Scheme of the snap-buckling mechanism in the hydrogel: **a** hydrogel in its pre-stressed state; **b** the hydrogel is wetted and expands, changing from a concave to a convex shape; **c** evaporation of the water, the pre-stressed geometry tries to turn back to a convex shape; **d** the hydrogel flips as the shape snaps back from convex to concave (adapted from Oliver et al. 2016)

To achieve a twisting movement, the two faces of the materials are reinforced so as to force the expansion into perpendicular directions during hydration, twisting to accommodate both pressures (Erb et al. 2013) (Fig. 5.13b).

Scientists developed a "jumping hydrogel" making use of a bi-stable snap-buckling geometry and built-in strain to self-propel the microgel into the air. This actuation mechanism combines a geometric activation mechanism—a shape and curvature deformation resulting from a snap-buckling instability (see ▶ Chap. 3, Sect. 3.5.2)—and an external trigger. The driving force in the snapping is the elastic curvature energy stored in the curved pressurized material, loaded like a spring. When further stimulated, stability drops drastically and, in order to regain equilibrium, the material snaps from a convex to a concave geometry (Persiani 2018). In the case of the hydrogel, the material is pre-stressed into a concave shape. As swelling is induced in the material through adding of water, the geometry elastically bends into a convex shape. When the water evaporates, equilibrium drops and the gel snaps back into the concave shape, propelling the mechanism into the air (Oliver et al. 2016) (Fig. 5.14).

5.3.4.3 Shape-Change Carbon Fibre

Carbon fibre composites can be combined with materials with shape-changing properties, to enhance the proprieties of the material (see also ▶ Sect. 5.2.3.5). By embedding humidity-activated SCM within the carbon fibre's microstructure, specific dynamic effects can be achieved.

Programmed carbon fibre surfaces and structures that react to specific humidity conditions are particularly interesting in the automotive industry to develop stabilizing geometries and develop safety measures (airfoils that can adapt to driving in different weather conditions) (MIT Self-assembly Lab 2019).

5.3.5 Natural Hygroscopic Materials

Shape-changing materials are widespread in nature and particularly in plants which have evolved mechanisms that rely on their internal heterogeneous architecture to achieve shape change upon external stimulus.

Most natural materials rely on localized directional and dimensional changes in the material to perform a wide range of movements (opening, bending, twisting and curling). The transmission and amplification of small degrees of strain across length are another effective way of translating relatively simple and modest response to external stimuli into large-scale movements.

A wide range of natural fibres can be distinguished by their origin, whether plant, animal or mineral. Naturally occurring mineral fibres are represented by asbestos, which are mostly water-repellent as well as a health hazard and will therefore not be further explored.

5.3.5.1 Hygroscopic Plant Fibres

Plants are able to deploy surprisingly fast and complex actuation mechanisms by exploiting the additive effect of simple volume changes and cellulose orientation driven by moisture uptake.

Plant fibres are composite structures made of an amorphous hemicellulose and lignin matrix with a cellulose microfibril reinforcement around the axis of the fibre. The hydrophilic behaviour of these fibres shows an unstable dimensional behaviour that depends on their composition (absorption capacity) and their specific structure. Typically, cellulose fibrils swell or shrink with some delay after the sorption phenomenon has taken place, and not instantly (Célino et al. 2014). As the anisotropic material structure finally swells, internal stresses form between the swelling fibres resulting in a pre-programmed structural and directional elongation.

To achieve faster responses through the sudden release of energy, these volume changes are often combined with other structural features such as cavitation bubbles used by fern spores or elastic potential energy in the leaves of the Venus flytrap (Persiani 2018).

Seedpods of orchid trees (sp. *Bauhinia variegata*) split open in two helical strips when drying. Each halve of the seedpod is composed of a bilayer structure with each layer expanding perpendicular to the other causing a frustrated geometry which twists as a result (Erb et al. 2013).

Pinecones (cl. *Pinopsida*) use a reversible hygroscopic actuation for opening and closing during seed dispersal. The movement is driven by structural differences in the orientation of the cellulose microfibrils within cell walls of the scales. The scales work as a bicomposite material, achieving a greater elongation on one side, while the other resists elongation, resulting in the scales opening (Dunlop et al. 2011) (◘ Fig. 5.15).

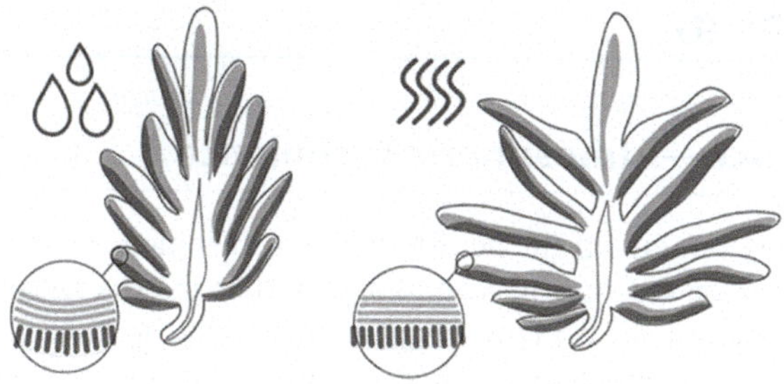

Fig. 5.15 Scheme of the bicomposite actuation of the scales in a pinecone (adapted from Dunlop et al. 2011)

Innovative approaches using natural cellulose fibres as wood grains to 3D print customized geometries are able to achieve a wide range of material transformations. The printing technique allows to selectively apply the cellulose material where needed to pre-programme specific structural capacities and direct the kinetic transformation (Correa et al. 2015). These materials are still under development; however, potential applications include self-assembling structures and packaging (Grönquist et al. 2019).

Other potential material developments include the layering of plant fibres and the combination with other natural or artificial materials to form bicomposite hygroscopic materials (Wood et al. 2018).

5.3.5.2 Hygroscopic Animal Fibres

A great number of animal fibres exhibit shape-memory behaviour due to their keratin content. This is the case of birds' feathers and human hair, which can be plastically deformed under warm temperature conditions (as with a curling iron) and recover their original shape when wetted (Oliver et al. 2016).

Human hair is specifically sensitive to humid air which contains a high number of water molecules, forming hydrogen bonds between the keratin proteins in hair that causes the curling. Because of its hygroscopic proprieties, applications include the use of human or animal hair as measuring mechanism in mechanical hygrometers (Encyclopædia Britannica 2019).

5.3.5.3 Biocomposites

Composite collagen non-woven networks can be realized by combining natural collagen fibres recycled from animal skin (SCF) within a polyurethane elastomeric matrix (PU) into a hygroscopic-activated shape-memory composite (SCF/PU). The water particles work as the switch between shape deformation and fixation by stimulating the destruction or the bonding of the collagen fibres. Recovery of the initial shape is achieved through the elastic nature of the matrix. The composite is biodegradable and has better biocompatibility than pure polyurethane (Han et al. 2019). Potential applications are mostly in the field of biomedicine (sensors, regenerative medicine, artificial skins).

5.4 Kinetic Energy

5.4.1 Turbulence—Atmospheric Turbulence

The kinetic energy contained in the atmospheric turbulence can be mainly used as direct actuator (see ▶ Chap. 4, Sect. 4.4.3). Other factors within the turbulent environment might influence materials also thermally or in terms of humidity. The potential for a combined actuation of multiple factors is therefore also possible (see ▶ Sect. 5.5).

5.4.2 Precipitations

Precipitations can trigger an autoreactive material mainly in terms of humidity change (see ▶ Sect. 5.3.4), but also as direct kinetic interaction (see ▶ Chap. 4, Sect. 4.4.3). The potential for a combined actuation of multiple factors is therefore also possible (see ▶ Sect. 5.5).

5.4.3 Direct Interaction—Applied Force

The transformation of kinetic energy into a specific kinetic movement can be achieved by (i) assembling several parts to form a mechanism to further transmit or amplify motion, or by (ii) employing the different elastic and plastic mechanical properties of the materials. The mechanisms referred to in the first case are discussed in a dedicated chapter (refer to ▶ Chap. 3). The mechanical behaviour of commonly used materials, referred to in the second case, is considered to be part of the state-of-the-art knowledge in the field of mechanics and dynamics and will not be further explored. A few uncommon and counter-intuitive material proprieties are however further discussed.

Mechanical metamaterials (MMs) are artificial structures with engineered mechanical properties tailored by intervening on their structure rather than on their chemical composition. MMs exhibit unique and tailorable properties such as negative Poisson's ratio, vanishing stiffness and negative compressibility (Surjadi et al. 2019).

5.4.3.1 Auxetic Materials

Auxetic materials show a counter-intuitive behaviour when subjected to mechanical stress or strain, due to the particular deformation mechanism of their internal structure which results in a negative Poisson's ratio. When stretched in the longitudinal direction, instead of thinning in the perpendicular direction to the stretching as most other materials, auxetic materials become thicker in one or several of the perpendicular width-wise directions. Similarly, when compressed in the

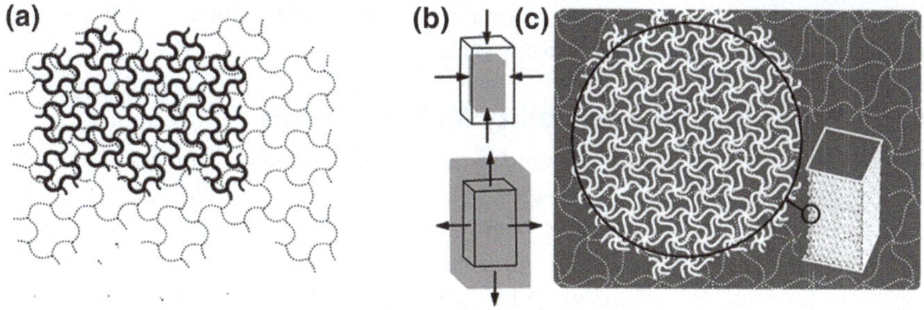

Fig. 5.16 Scheme of typical auxetic geometries: **a** 2D auxetic geometry; **b** behaviour of the auxetic material when compressed and expanded; and **c** scheme of an auxetic material structure on microscopic scale

longitudinal direction, the material becomes thinner in transverse dimensions (see also ▶ Chap. 2). Partially auxetic materials reveal an auxetic behaviour in some directions and not in others, as in α-cristobalite (Carneiro et al. 2013).

Auxetic proprieties can be possessed by material structures both on a microscopic and on a macroscopic level, embedded in the material or in the geometry of the sample. Auxetic cellular structures often have re-entrant, chiral or rotating geometries that enable the mechanical shrinkage and expansion of the cells constituting the material (Papadopoulou et al. 2017) (Fig. 5.16).

Possible auxetic geometries are numerous and include, among others, systems using squares, rectangles, triangles and other geometries. Additional models have been formulated hypothesizing that the units are flexible (Barchiesi et al. 2019).

Specific proprieties possessed by these material structures that make them interesting design partners are (Carneiro et al. 2013):

- Indentation resistance;
- High energy absorption;
- Fracture resistance;
- High resistance to shear;
- Acoustic absorption;
- Variable permeability;
- Synclastic behaviour (deform in a dome shape when bent);
- Geometric control of the structure.

Typical auxetic materials are certain polyurethane (PU) foams, certain rocks and minerals, polytetrafluoroethylene polymers (Gore-Tex) (Grima and Evans 2006), tendons (Gatt et al. 2015), paper (Verma et al. 2013), chain organic molecules (n-paraffins), cork. Auxetic structures include non-woven textiles as yarns/stitches/knit loops, origami folds like the diamond-folding-structure (RFS), the herringbone-fold-structure (FFS) or the Miura fold (see ▶ Chap. 2).

Applications for materials with these proprieties are mostly those that require specific mechanical properties such as high energy absorption and fracture resistance: defence and impact absorption (military, sports, automotive, etc.), wearables (comforting to several body shapes and long wear fabrics), medical (drug dispensers, wound care, compression garments).

5.4.3.2 Negative Compressibility Metamaterials

Most conventional materials expand when tensioned and contract when compressed, regardless of how they respond in transverse directions, deforming along the direction of the applied force. Metamaterials with negative longitudinal and volume compressibility are similar to auxetic materials in that they are designed to achieve the opposite behaviour. They undergo contraction when tensioned and expansion when compressed, but always along the direction of the applied force.

Negative compressibility, as an inherent material property, is extremely uncommon as it is thermodynamically unfavourable. Apart from methanol monohydrate, a simple molecular crystal with negative linear compressibility and negative and anisotropic thermal expansion, most ways of obtaining these metamaterials has been to design specific micromechanisms achieving negative compressibility (Barchiesi et al. 2019).

These materials have a potential use as artificial muscles, actuators, force amplifiers, micromechanical controls, pressure sensors and protective devices (Surjadi et al. 2019).

5.4.3.3 Pentamode Metamaterials

Pentamode metamaterials, also known as metafluids, are three-dimensional solids with high bulk modulus and small shear modulus, similar to an ideal fluid. Metafluids undergo deformation at a constant volume: the material is hard to compress yet easy to deform in five out of the six principal directions[1] (Barchiesi et al. 2019).

These materials are still in their early phases of development but have potential applications as wearables ("unfeelable" cloak), in the medical field (tissue regeneration performance), acoustics (Hedayati et al. 2017).

5.4.3.4 Elastic Chiral Metamaterial

Elastic chiral metamaterials have an internal reflection symmetry allowing them to respond to compression with a twisting motion.

Chirality is a specific property of asymmetry which makes it distinguishable from its mirror image as the geometry cannot be superimposed on it by rotations and translations (Frenzel et al. 2017). Chiral materials are built around central nodes that can vary from a circular to a rectangular shape with many possible shapes in between that are then joined together through straight ligaments (Barchiesi et al. 2019) (◘ Fig. 5.17).

---

1 The six directions of deformation of a body in a three-dimensional coordinate system can be described by defining the axis along which the deformation occurs (x, y, z) and the direction of deformation (+; −).

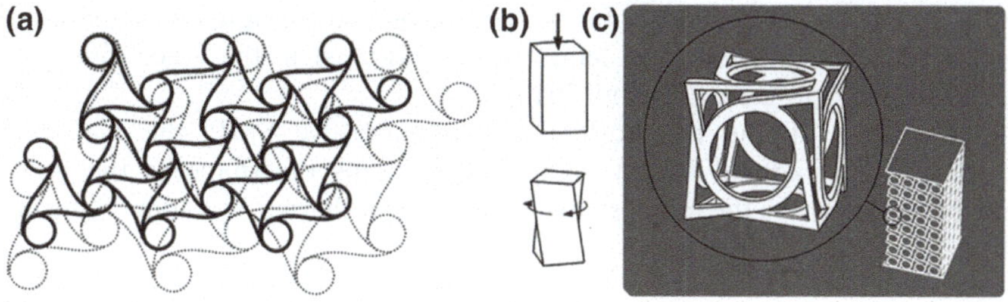

Fig. 5.17 Scheme of two different chiral structures. **a** 2D chiral geometry with auxetic behaviour (adapted from Barchiesi et al. 2019); **b** behaviour of the chiral material when compressed and twisted; and **c** 3D chiral geometry (adapted from Frenzel et al. 2017)

5.5 Multi-stimuli-Responsive Materials

Multi-stimuli-responsive materials combine the proprieties of two or more of the materials mentioned in this chapter to enable triggering by multiple energy sources with different dynamic outcomes.

Most multi-responsive materials are generated as composites, adding and connecting single stimulus-responsive materials. Few are however being developed as inherently multi-responsive (Yang et al. 2017). Moreover, although these materials have multiple potential applications, many need to be reprogrammed before being able to respond to other stimuli (Xie et al. 2017).

As the possibilities are multiple and new combinations are constantly developed, only a few established materials will be discussed.

5.5.1 Multi-responsive Shape-Memory Polymers

Multi-responsive shape-memory polymers allow to combine several stimuli and unique properties with dual-responsive (thermo/solvent, thermo/water, thermo/photo) and triple-responsive shape-memory effects (thermo-/photo-/chemo-responsive) where the polymer chains undergo reversible alteration in structure and morphology upon exposure to multiple triggers (Yang et al. 2017; Kumpfer and Rowan 2011).

Thermo-/moisture-responsive SMPs are able to recover their original shape when immersed in water, as well as at a specific temperature. The main advantage of this system is that the moisture-driven shape recovery is able to significantly lower down the glass transition temperature of the polymer, allowing the material to react long before the triggering temperature is reached. This allows the same material to be operated with different transition temperatures at different locations (Huang et al. 2010).

In another example of a thermo-/moisture SMP, a PNIPAAM–alumina hydrogel was manufactured as a bilayer composite with programmed orientation-dependent stiffness on microstructural level within a swellable/shrinkable matrix. The hydrogel would be able to twist in one direction upon increasing the temperature over the transition temperature and twist in the other direction if allowed to swell in water (Erb et al. 2013).

In the case of the thermo-/photo-dual-responsive SMP, the material is deformed when exposed to UV light, and it keeps the shape as the light is removed and recovers the initial shape by applying a thermal stimulus (Yang et al. 2017).

5.5.2 Active Auxetics

Active auxetic materials combine the temporary kinetic proprieties of SMMs with the dynamic geometric proprieties of auxetic materials. The combination possibilities are multiple: light-/heat-/humidity-active auxetics. What all these variations have in common is that the geometric dynamic potential of the auxetic geometry is autonomously activated, and in case of materials that expand or swell in one or more dimensions, the overall expanding effect can be multiplied through the combined action of both systems.

Active auxetics are therefore important tools in design applications, offering flexibility through precise and repeatable behaviour as well as functional and environmental adaptability (Papadopoulou et al. 2017).

5.6 Conclusion

Material scientists are now able to increasingly programme complex intrinsic behaviour in materials, from high-precision and repeatable motion geometries to functional and environmental adaptability. We can design materials to be actuated and move in a predetermined and customizable sequence, opening up to new possibilities to drive our artefacts, the materials being the machines.

This chapter has defined the range of materials that can be considered autoreactive or that can be used in an autoreactive context. It has explored the broad possibilities, available in the current state of the art, to drive kinetic actuation through the use of different latent energy sources with the materials as the vehicle of the energetic transformation. Materials have been classified according to the source of the energetic input and then according to their chemical and physical properties.

The potential impact of these materials on future technological innovations is anticipated to be huge, changing the way we relate to our surroundings with corresponding benefits in quality of life. The possibility to embed multifunctional and more complex behaviours in objects could ideally reduce the number of tools and components, unlocking applications, uses and designs not yet considered. Biocompatible and biodegradable functional alternatives are becoming increasingly available, being even in some cases sustainably and renewably grown from a manufacturing perspective with potential environmental benefits.

On the other hand, the main question of the necessity of available latent energy sources corresponding to the wished outcomes and to the timing for autoreaction remains an intrinsic limit of these systems that reduce the range of possible applications. Also, although these materials become increasingly market-present and economically available, many of them are still out of range for a multitude of everyday applications and in less progressive fields as, for instance, in the building industry.

As the final aim with designing autoreactive mechanisms is to optimize the energetic output to input, we strive to create structures that minimize the force required to self-shape. This can be achieved in multiple ways: targeting the combined use of complementary stimuli to maximize the magnitude of the trigger delivered, using combined structural and chemical techniques to overcome the shortcomings, embedding features crossing multiple length scales, copying and borrowing from solutions developed by nature.

It is therefore important for designers and engineers to keep in mind the many possibilities to achieve autoreactivity, not forgetting the potential for synergy between materials and form that in some cases can achieve good autoreactive design without the use of autoreactive materials, but through simple and smart design solutions.

References

Arun DI, Chakravarthy P, Arockia Kumar R, Santhosh B (2018) Shape memory materials. CRC Press, Boca Raton. ► https://doi.org/10.1201/9781351119948

Balmer A, Martin P (2008) Synthetic biology, social and ethical challenges. An independent review commissioned by the Biotechnology and Biological Sciences Research Council (BBSRC). Retrieved on 28 Oct 2019 from ► http://www.haseloff-lab.org/resources/SynBio_reports/0806_synthetic_biology.pdf

Barchiesi E, Spagnuolo M, Placidi L (2019) Mechanical metamaterials: a state of the art. Math Mech Solids 24(1):212–234. ► https://doi.org/10.1177/1081286517735695

Benyus J (2002) Biomimicry: innovation inspired by nature. Harper Perennial, New York

Brownell B (2005) Transmaterial: a catalog of materials that redefine our physical environment. Princeton Architectural Press, New York

Buffington SL, Paula JE, Aliabc MM, Maciosa MM, Mather PT, Henderson JH (2019) Enzymatically triggered shape memory polymers. Acta Biomater 84:88–97. ► https://doi.org/10.1016/j.actbio.2018.11.031

Callaway E (2019) CRISPR cuts turn gels into biological watchdogs. Nature 572:574. ► https://doi.org/10.1038/d41586-019-02542-3

Carneiro VH, Meireles J, Puga H (2013) Auxetic materials—a review. Mater Sci-Poland 31(4):561–571. ► https://doi.org/10.2478/s13536-013-0140-6

Célino A, Fréour S, Jacquemin F, Casari P (2014) The hygroscopic behavior of plant fibers: a review. Front Chem 1:43. ► https://doi.org/10.3389/fchem.2013.00043

Chen S, Mo F, Chen S, Ge Z, Yang H, Zuo J, Liu X, Zhuo H (2015) New insights into multi-shape memory behaviours and liquid crystalline properties of supramolecular polyurethane complexes based on pyridine-containing polyurethane and 4-octyldecyloxybenzoic acid. J Mater Chem A 3(38):19525–19538. ► https://doi.org/10.1039/C5TA04469C

Clarke AC (1984) Profiles of the future: an inquiry into the limits of the possible. Holt, Rinehart, and Winston. ISBN 9780030697838

Collins JJ, Maxon M, Ellington A, Fussenegger M, Weiss R, Sauro H (2014) Synthetic biology: how best to build a cell. Nat News Comment 509(7499):155–157. ► https://doi.org/10.1038/509155a

Correa D, Papadopoulou A, Guberan C, Jhaveri N, Reichert S, Menges A, Tibbits S (2015) 3D-printed wood: programming hygroscopic material transformations. 3D Print Addit Manuf 2(3). ► https://doi.org/10.1089/3dp.2015.0022

Darvishmanesh S, Qian S, Wickramasinghe SR (2015) Responsive membranes. In: Schneider HJ (ed) Chemoresponsive materials: stimulation by chemical and biological signals. Royal Society of Chemistry. ► https://doi.org/10.1039/9781782622420

Dunlop JWC, Weinkamer R, Fratzl P (2011) Artful interfaces within biological materials. Mater Today 14(3):70–78. ► https://doi.org/10.1016/S1369-7021(11)70056-6

Encyclopædia Britannica (2019) Hygrometer. Encyclopædia Britannica, Inc. Published 18 Dec 2019. Retrieved on 22 Apr 2020 from ► https://www.britannica.com/science/hygrometer

Erb R, Sander J, Grisch R, Studart AR (2013) Self-shaping composites with programmable bioinspired microstructures. Nat Commun 4:1712. ► https://doi.org/10.1038/ncomms2666

Fox MA, Kemp M (2009) Interactive architecture. Princeton Architectural Press, New York. ISBN 978-1568988368

Frenzel T, Kadic M, Wegener M (2017) Three-dimensional mechanical metamaterials with a twist. Science 358(6366):1072–1074. ► https://doi.org/10.1126/science.aao4640

Gatt R, Wood MV, Gatt A, Zarb F, Formosa C, Azzopardi KM, Casha A, Agius TP, Schembri-Wismayer P, Attard L, Chockalingam N, Grimaah JN (2015) Negative Poisson's ratios in tendons: an unexpected mechanical response. Acta Biomater 24:201–208. ► https://doi.org/10.1016/j.actbio.2015.06.018

Ge Q, Sakhaei AH, Lee H, Dunn CK, Fang NX, Dunn ML (2016) Multimaterial 4D printing with tailorable shape memory polymers. Sci Rep 6:31110. ► https://doi.org/10.1038/srep31110

G-RAU (2015) ► https://www.g-rau.de/fileadmin/gustavrau/downloads/service/downloads/thermal_actuators.pdf

Grima JN, Evans KE (2006) Auxetic behavior from rotating triangles. J Mater Sci 41(10):3193–3196. ► https://doi.org/10.1007/s10853-006-6339-8

Grönquist P, Wood D, Hassani MM, Wittel FK, Menges A, Rüggeberg M (2019) Analysis of hygroscopic self-shaping wood at large scale for curved mass timber structures. Sci Adv 5(9), eaax1311. ► https://doi.org/10.1126/sciadv.aax1311

Gruber P (2011) Biomimetics in architecture: architecture of life and buildings. Springer-Verlag, Wien. ISBN 978-3-7091-0331-9

Han D, Lu Z, Chester SA, Lee H (2018) Micro 3D printing of a temperature-responsive hydrogel using projection micro-stereolithography. Sci Rep 8:1963. ► https://doi.org/10.1038/s41598-018-20385-2

Han Y, Hu J, Chen X (2019) A skin inspired bio-smart composite with water responsive shape memory ability. Mater Chem Front 3:1128–1138. ► https://doi.org/10.1039/C9QM00114J

Havens E, Snyder EA, Tong TH (2005) Light-activated shape memory polymers and associated applications. In: Proceedings SPIE. Smart structures and materials 2005: industrial and commercial applications of smart structures technologies, vol 5762, p 48. ► https://doi.org/10.1117/12.606109

He B, Li H, Jin K (2013) Shape memory polymer actuated hollow snap-fit design analysis. Mater Des 47:539–550. ► https://doi.org/10.1016/j.matdes.2012.12.038

Hedayati R, Leeflang AM, Zadpoor AA (2017) Additively manufactured metallic pentamode meta-materials. Appl Phys Lett 110:091905. ► https://doi.org/10.1063/1.4977561

Huang WM, Yang B, Zhao Y, Ding Z (2010) Thermo-moisture responsive polyurethane shape-memorypolymer and composites: a review. J Mater Chem 20:3367–3381. ► https://doi.org/10.1039/B922943D

Iqbal D, Samiullah MH (2013) Photo-responsive shape-memory and shape-changing liquid-crystal polymer networks. Materials 6(1):116–142. ► https://doi.org/10.3390/ma6010116

Janocha H (ed) (1992) Aktoren, Grundlagen und Andwendungen. Springer-Verlag, Berlin. ISBN 978-3-662-00418-0

Janocha H (ed) (2007) Adaptronics and smart structures: basics, materials, design, and applications, 2nd edn. Springer-Verlag, Berlin. ISBN 978-3-540-71965-6

Kan V, Vargo E, Machover N, Ishii H, Pan S, Chen W, Kakehi Y (2017) Organic primitives: synthesis and design of pH-reactive materials using molecular I/O for sensing, actuation, and interaction. In: CHI'17 proceedings of the 2017 CHI conference on human factors in computing systems, Denver, 06–11 May 2017, pp 989–1000. ► https://doi.org/10.1145/3025453.3025952

Kannus P (2000) Structure of the tendon connective tissue. Scand J Med Sci Sports 10:312–320. ► https://doi.org/10.1034/j.1600-0838.2000.010006312.x

Kim YS, Liu M, Ishida Y, Ebina Y, Osada M, Sasaki T, Hikima T, Takata M, Aida T (2015) Thermo-responsive actuation enabled by permittivity switching in an electrostatically anisotropic hydrogel. Nat Mater 14:1002–1007. ► https://doi.org/10.1038/nmat4363

Kirillova A, Ionov L (2019) Shape-changing polymers for biomedical applications. J Mater Chem B 7:1597–1624. ► https://doi.org/10.1039/C8TB02579G

Kohl M (2004) Shape memory microactuators. Springer-Verlag, Berlin. ISBN 3-540-20635-3

Kumpfer JR, Rowan SJ (2011) Thermo-, photo-, and chemo-responsive shape-memory properties from photo-cross-linked metallo-supramolecular polymers. J Am Chem Soc 133(32):12866–12874. ► https://doi.org/10.1021/ja205332w

Lai A, Du Z, Gan CL, Schuh CA (2013) Shape memory and superelastic ceramics at small scales. Science 341(6153):1505–1508. ► https://doi.org/10.1126/science.1239745

Lendlein A (ed) (2010) Shape-memory polymers. Springer-Verlag, Berlin. ISBN 978-3-642-12358-0

Lendlein A, Kelch S, Kratz K, Schulte J (2005) Shape-memory polymers. In: Encyclopedia of materials: science and technology, 2nd edn. Elsevier Ltd., pp 1–9

Luzzati V, Mustacchi H, Skoulios A (1957) Structure of the liquid-crystal phases of the soap–water system: middle soap and neat soap. Nature 180:600–601

Maeda S, Hara Y, Yoshida R, Hashimoto S (2007) Chemical robot—design of self-walking gel. In: IEEE/RSJ international conference on intelligent robots and systems, 29 Oct–2 Nov 2007, San Diego. ► https://doi.org/10.1109/iros.2007.4399392

Maeda S, Hara Y, Nakamaru S, Nakagawa H, Hashimoto S (2011) Chemical robots. In: Pramatarova LD (ed) On biomimetics. IntechOpen. ► https://doi.org/10.5772/18342. Available from: ► https://www.intechopen.com/books/on-biomimetics/chemical-robots

Maerk SR, Gran ML, Peppas NA, Caldorera-Moore M (2015) Intelligent, responsive and teranostic hydrogel systems for controlled delivery of therapeutics. In: Schneider HJ (ed) Chemoresponsive materials: stimulation by chemical and biological signals. Royal Society of Chemistry. ► https://doi.org/10.1039/9781782622420

Menges A, Hensel M (2008) Morpho-ecologies. Architectural Association, London. ISBN 978-1902902531

Miller W, Smith CW, Mackenzie DS, Evans KE (2009) Negative thermal expansion: a review. J Mater Sci 44:5441–5451. ► https://doi.org/10.1007/s10853-009-3692-4

MIT Self-assembly Lab (2019) Programmable materials. Retrieved 6 Nov 2019 from ► https://selfassemblylab.mit.edu/programmable-materials

Miyata T (2015) Target biomolecule-responsive hydrogels. In: Schneider HJ (ed) Chemoresponsive materials: stimulation by chemical and biological signals. Royal Society of Chemistry. ► https://doi.org/10.1039/9781782622420

Nachtigall W (2005) Biologisches Design, Systematischer Katalog für bionisches Gestalten. Springer Verlag, Berlin. ISBN 978-3540227892

Nakata M, Zanchetta G, Chapman BD, Jones CD, Cross JO, Pindak R, Bellini T, Clark NA (2007) End-to-end stacking and liquid crystal condensation of 6– to 20–base pair DNA duplexes. Science 318(5854):1276–1279. ► https://doi.org/10.1126/science.1143826

Oliver K, Seddon A, Trask RS (2016) Morphing in nature and beyond: a review of natural and synthetic shape-changing materials and mechanisms. J Mater Sci 51(24):10663–10689. ► https://doi.org/10.1007/s10853-016-0295-8

Papadopoulou A, Laucks J, Tibbits S (2017) Auxetic materials in design and architecture. Nat Rev Mater 2:17078. ► https://doi.org/10.1038/natrevmats.2017.78

Persiani SGL (2018) Biomimetics of motion: nature-inspired parameters and schemes for kinetic design. Springer International Publishing, Cham. ► https://doi.org/10.1007/978-3-319-93079-4

Persiani SGL, Battisti A (2018) Frontiers of adaptive design, synthetic biology and growing skins for ephemeral hybrid structures. In: Energy-efficient approaches in industrial applications. IntechOpen. ▶ https://doi.org/10.5772/intechopen.80867

Persiani SGL, Molter PL, Aresta C, Klein T (2016) Mapping of environmental interaction and adaptive materials for the autoreactive potential of building skins. In: Proceedings of the 41st IAHS world congress on sustainability and innovation for the future, Albufeira. ISBN 978-989-98949-4-5

Petrović ZS, Milić J, Zhang F, Ilavsky J (2017) Fast-responding bio-based shape memory thermoplastic polyurethanes. Polymer 121:26–37. ▶ https://doi.org/10.1016/j.polymer.2017.05.072

Picado J (2010) Seashells: the plainness and beauty of their mathematical description. Mathematical Association of America. Retrieved 3 Mar 2015 from ▶ http://www.maa.org/publications/periodicals/loci/seashells-the-plainness-and-beauty-of-their-mathematical-description

Ritter A (2007) Smart materials in architecture, interior architecture and design. Birkh.user, Basel. ISBN 13: 978-3-7643-7327-6

Santo L (2016) Shape memory polymer foams. Prog Aerosp Sci 81:60–65. ▶ https://doi.org/10.1016/j.paerosci.2015.12.003

Sauer C (2010) Made of… new materials sourcebook for architecture and design. Gestalten, Berlin

Sessini V, Brox D, López AJ, Ureña A, Peponi L (2018) Thermally activated shape memory behavior of copolymers based on ethylene reinforced with silica nanoparticles. Nanocomposites 4(2):19–35. ▶ https://doi.org/10.1080/20550324.2018.1472723

Surjadi JU, Gao L, Du H, Li X, Xiong X, Fang NX, Lu Y (2019) Mechanical metamaterials and their engineering applications. Adv Eng Mater 21(3):1800864. ▶ https://doi.org/10.1002/adem.201800864

Takenaka K (2012) Negative thermal expansion materials: technological key for control of thermal expansion. Sci Technol Adv Mater 13(1). ▶ https://doi.org/10.1088/1468-6996/13/1/013001

Testa P, Style RW, Cui J, Donnelly C, Borisova E, Derlet PM, Dufresne ER, Heyderman LJ (2019) Magnetically addressable shape-memory and stiffening in a composite elastomer. Adv Mater 31:1900561. ▶ https://doi.org/10.1002/adma.201900561

Verma P, Shofner ML, Griffin AC (2013) Deconstructing the auxetic behavior of paper. Phys Status Solidi 251(2):289–296. ▶ https://doi.org/10.1002/pssb.201384243

Vollrath F, Knight DP (2001) Liquid crystalline spinning of spider silk. Nature 410:541–548

Wan T (2010) Shape memory polymer yarns. In: Technical textile yarns. Woodhead Publishing Series in Textiles, pp 429–451. ▶ https://doi.org/10.1533/9781845699475.2.429

Wei ZG, Sandstrom R, Miyazaki S (1998) Shape memory materials and hybrid composites for smart systems: part II shape-memory hybrid composites. J Mater Sci 33:3763. ▶ https://doi.org/10.1023/A:1004674630156

Wojtowicz PJ (1975) Lyotropic liquid crystals and biological membranes: the crucial role of water. In: Priestley EB, Wojtowicz PJ, Sheng P (eds) Introduction to liquid crystals. Springer, Boston. ▶ https://doi.org/10.1007/978-1-4684-2175-0_18

Wood D, Vailati C, Menges A, Rüggeberg M (2018) Hygroscopically actuated wood elements for weather responsive and self-forming building parts—facilitating upscaling and complex shape changes. Constr Build Mater 165. ▶ https://doi.org/10.1016/s0950061817325394

Xiao F, Naficy S, Casillas G, Khan MH, Katkus T, Jiang L, Liu H, Li H, Huang Z (2015) Edge-hydroxylated boron nitride nanosheets as an effective additive to improve the thermal response of hydrogels. Adv Mater 27(44):7196–7203. ▶ https://doi.org/10.1002/adma.201502803

Xie H, Deng X-Y, Cheng C-Y, Yang K-K, Wang Y-Z (2017) New strategy to access dual-stimuli-responsive triple-shape-memory effect in a non-overlapping pattern. Macromol Rapid Commun 38:1600664. ▶ https://doi.org/10.1002/marc.201600664

Xie H, Shao J, Ma Y, Wang J, Huang H, Yang N, Wang H, Ruan C, Luo Y, Wang QQ, Chu PK, Yu XF (2018) Biodegradable near-infrared-photoresponsive shape memory implants based on black phosphorus nanofillers. Biomaterials 164:11–21. ▶ https://doi.org/10.1016/j.biomaterials.2018.02.040

Yamada M, Kondo M, Mamiya J, Yu Y, Kinoshita M, Barrett CJ, Ikeda T (2008) Photomobile polymer materials: towards light-driven plastic motors. Angew Chem Int Ed 47:4986–4988. ▶ https://doi.org/10.1002/anie.200800760

Yang J, Wen H, Zhuo H, Chen S, Ban J (2017) A new type of photo-thermo staged-responsive shape-memory polyurethanes network. Polymers 9(7), 287. ► https://doi.org/10.3390/polym9070287

Yu YL, Nakano M, Ikeda T (2003) Directed bending of a polymer film by light—miniaturizing a simple photomechanical system could expand its range of applications. Nature 425:145. ► https://doi.org/10.1038/425145a

Zhang LC, Wang L (2018) Microstructure and mechanical properties of beta type Ti-Fe based alloys. In: Development and application of biomedical titanium alloys. Bentham Science Publishers, pp 51–72. ► https://doi.org/10.2174/9781681086194118010005

Zhang L, Huang M, Yu R, Huang J, Dong X, Zhang R, Zhu J (2015) Bio-based shape memory polyurethanes (bio-SMPUs) with short side chains in the soft segment. J Mater Chem A 2:11490–11498. ► https://doi.org/10.1039/C4TA01640H

Design of Autoreaction, Case Studies

Contents

S. G. L. Persiani, *Design of Autoreaction*,
Design Science and Innovation, https://doi.org/10.1007/978-981-15-6178-8_6

The growing technical capacity to integrate living-like features in artefacts and products, such as automated homeostasis, reactivity and physical change, has fuelled much scientific, technical and artistic research in the last decade. This trend has also been supported by the progressive change of the production processes, shifting from the manufacturing of serial solutions towards one-of solutions. As a result, also the focus of designers is shifting from the optimization of the production processes towards the design of single solutions that make the most out of their shape and material use (Oosterhuis 2012). It is therefore logical to assume that in that context, the first fields where substantial innovation can be introduced, benefiting from the multidisciplinary knowledge of an interdisciplinary autoreactive approach, are design- and art-related fields.

The previous chapters have explored autoreactive parameters in very diverse contexts, with examples taken from many engineering contexts. This chapter aims to get a closer look at the autoreactive state of art in the fields of architecture, design and art through the analysis of the kinetic features of a diverse array of case studies.

Technical innovations are rarely developed in the field of architecture: solutions are traditionally built to last for long times and are applied over very big scales, prioritizing cautious and low-cost solutions (Persiani and Battisti 2018). Introducing moving elements can be risky as it implies wear and potentially high maintenance costs, which need to be justified by an added efficiency to the system. Hence, moving parts in architecture are mostly relatively simple with uniformed modules. They are big enough to minimize the number of moving parts, but also small enough to be managed by the users. Therefore, introducing responsive elements within buildings brings a true revolution to the whole sector and raises the necessity to rethink dogmatic preconceptions as longevity and stability along with the integration of optimized kinetic features (Kretzer and Hovestadt 2014). However, long before new concepts and materials are adopted in architecture, ideas are first experimented and developed as student projects, as installations, or are imported from closely related fields as design and art, where experiments can be brought further at lower risks.

Industrial design is going through huge changes due to the growing interest and demanding taste of consumers in terms of use of new materials and technologies. The insertion of kinetic features in design products is on one hand a way to charm the users, choreographing the physical movement and on the other hand a mean to invent and develop entirely new design concepts and markets. As ugliness does not sell and companies commit a lot of resources to design every aspect of a product, as much attention is given to the object's moving features as to its form, colour or material—making of the field of industrial design a potential prosperous field where new autoreactive solutions and applications can be developed.

In comparison with industrial design, the field of art allows an even faster experimentation of new concepts, being free from most user's needs and requirements as well as from regulation. This allows art to remain more playful and achieve faster and more varied design experimentation than any other field. Artworks can be fairly inexpensive, fast to realize and also lighter to move. This means that they can potentially be exhibited in non-standard locations as well as tested on a broader category of users and are therefore a great occasion to experiment with isolated ideas and concepts, advertising itself and raising the interest of users and developers (Fox and Kemp 2009). It is not a chance that many new solutions that have further developed in architecture have started as part of an artwork or an exhibition pavilion.

The following analysis discusses the autoreactive potential of 27 different projects with geometrical or mechanical features of particular interest. Out of these, some integrate autoreactive principles while others have the potential to be redesigned to become autoreactive. For an easier comparison, the examples are further classified into five different macro-categories highlighting the principal mechanical systems characterizing their type of motion.

6.1 Scissor Structures and Planar Linkages

Scissor structures are mostly made out of two rigid or semi-flexible members which can be opened and closed: a mobile part is pressed against a rigid outgrowth or both parts are mobile and specular, meeting on the central axis. In nature, scissor structures are mostly used in pincers or beaks and mandibles, and have therefore often sharp edges, bulges or knobs to prevent the grip from slipping (Persiani 2018).

In architectural engineering, scissor structures are often used for mobile and temporary applications. The scissor mechanism is used as a low-tech and durable solution with both kinematic (during the deploying phase) and load-bearing proprieties (as a skeleton). The kinetic proprieties are strongly linked to the geometry of the system.

6.1.1 Solstice Clock

The Solstice Clock is a kinetic scissor structure (Gilbert 2018). The movement of the clock arms with the passing of time is used to drive a circular scissor structure that frames the clock. As the hours pass, the scissor structure expands and shrinks, switching between intermediate geometries, from a semi-circular shape as the arm strikes 12, to a 12-pointed flower shape as the arm strikes 6 (◘ Fig. 6.1). The speed of the expanding and shrinking movement of the scissor structure is quite slow, being under the range of velocity perceived by the human eye.

In the Solstice Clock, the powered mechanism is the motion of the clock arms, which has the primary and independent purpose to track time. The kinetic scissor structure adds to the mechanism of the regular clock a secondary purpose, which is to represent an abstract expression of time. The scissor structure can therefore be considered as *autoreactive* in energetic terms as it uses an existing movement with a different purpose to achieve kinetic change.

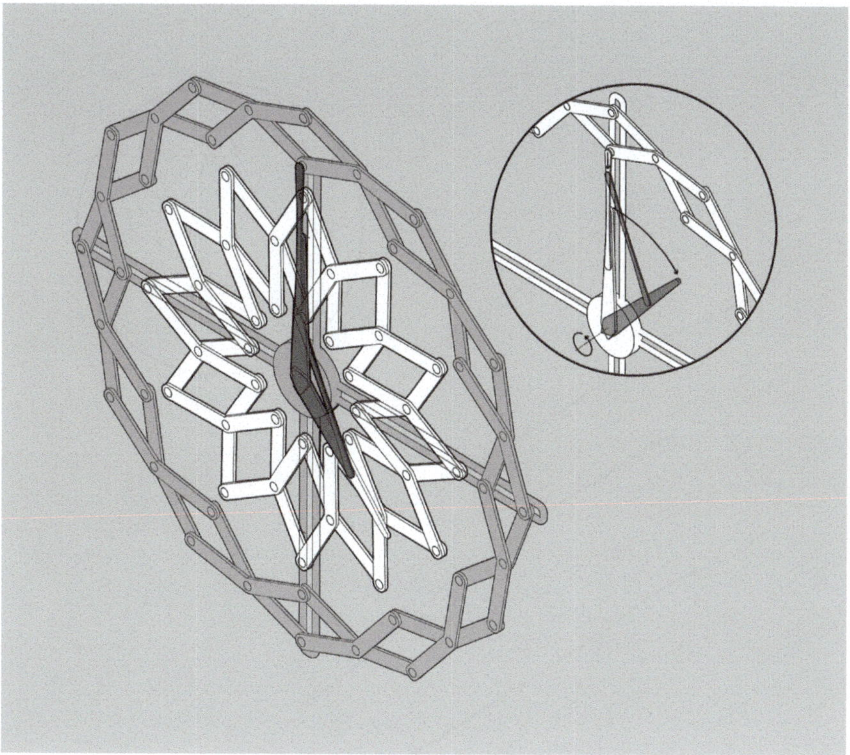

Fig. 6.1 Schematic representation of the functioning of the Solstice Clock (Kobas and Koth 2020) showing two different stages in the scissor frame's movement (left) and the detail of the linkage between the frame and the clock arm which drives the movement (right)

6.1.2 Polymorphic Bench

The polymorphic bench (Able et al. 2011) is a kinetic installation and piece of urban furniture built as a succession of 119 connected see-saw sections in plywood. The structure is connected through a pivot and sliding bolt system that transmits the vertical movement of one section to the neighbouring sections, while a series of internal elastic notches provide lateral stability during motion. The reactive shape opens up to new means of interaction, encouraging users to climb inside, lounge, jump, push and rock the installation. The result is halfway between a social chaise longue and an interactive balance board, as the undulating seating landscape is transformed under the direct action of the weight of its users (Fig. 6.2).

In energetic terms, the work necessary to achieve the shape changes of the bench needs to be directly provided by the users. The system therefore corresponds more to a *reactive* system rather than to an autoreactive one.

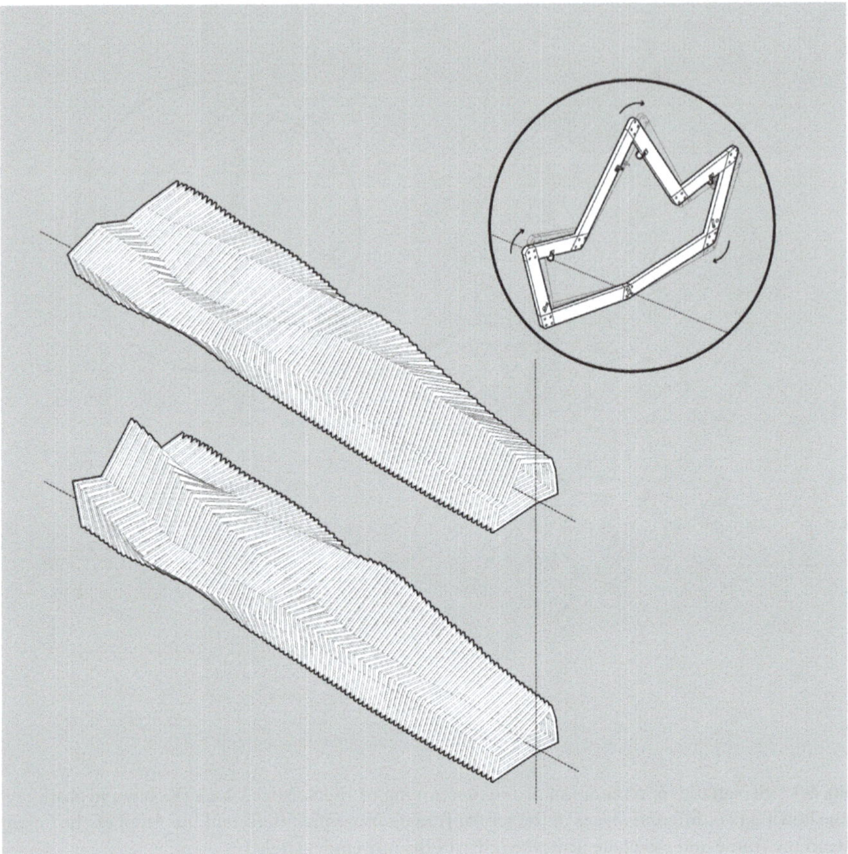

Fig. 6.2 Schematic representation of the polymorphic bench (Kobas and Koth 2020) with two of the many configurations of the overall bench (left) and in detail, the typical see-saw section of the seating (right)

6.1.3 Remembrane

The Remembrane flexible scissor structure system (Jun et al. 2015) was originally designed to drive a mechanical ventilation opening system to be installed within a facade. The special mechanics of the structure, developed combining the mechanics of a pantograph and a tensegrity structure (Fu 2018), compress or extend using a rhomboidal scissor pattern that enables a huge kinetic reaction with a small actuating movement (Fig. 6.3).

The structure is regulated by a sensor-actuator system measuring the changes in the environment in terms of wind pressure, humidity, temperature and CO_2 content. The scissor structure integrates nitinol wires and springs, actuated through the application of an electrical current. These wires, coiled up as springs and integrated in the structure, act as muscles lengthening and shortening the spacing between the rigid elements. Depending on the stimulation of the wires, the system can take intermediate positions in the opening/closing process: the structure can be made to bend in all directions, shorten and stretch out.

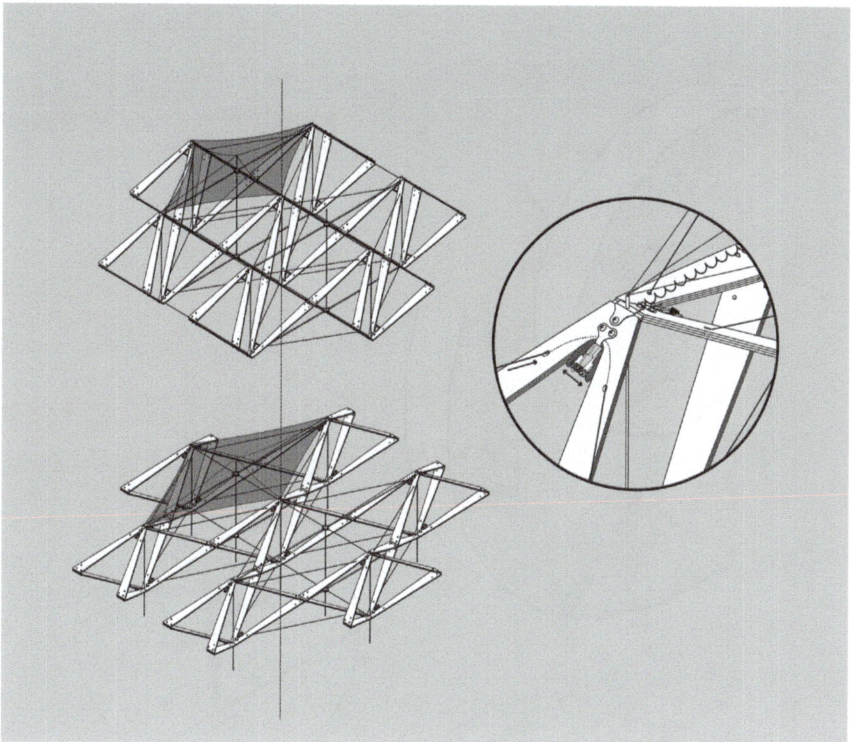

Fig. 6.3 Schematic representation of the kinetics performed in the Remembrane system (Kobas and Koth 2020). Compression and extension of the structure is schematized (left) and a detail of the linkage between the rigid parts showing the movement allowed and the connection to the NiTi wiring (right)

The characters of this system correspond mostly to those of an *adaptronic* system, being controlled through a combination of a sensor-brain system and a functional material that enables the actuation. Although not being autoreactive, this type of structure can be of good use in the design of autoreactive components. The system itself could be made autoreactive by substituting the electricity-triggered wires by temperature-triggered shape-memory wires. To closely control the bending and stretching movements of the system, the wiring system would however need to be designed with more detail.

6.1.4 Rolling Bridge

The Rolling Bridge at Paddington Basin (Heatherwick 2002) spans across a 12-meter-wide inlet of the Grand Union Canal like a conventional steel and timber footbridge which in this case rolls up on the bank of the canal to allow the passage of the boats, mimicking the curling of a caterpillar (Fig. 6.4).

Eight triangular steel sections are mounted intermittently with vertical hydraulic cylinders concealed within the vertical posts of the parapets. As the cyl-

Fig. 6.4 Kinetic functioning scheme of the Rolling Bridge (Kobas and Koth 2020): rolling and unrolling geometric stages (left) and detail of the hydraulic cylinders driving the movement from the parapets (right)

inders are simultaneously actuated, lifting the parapets, the triangular steel sections fold and curl, forming an octagonal shape as the ends join.

Due to its main function, which requires flexibility and reaction timeframes that are very unpredictable (boats passing through the canal), this system has been realized using a typical mechanically driven *responsive* system. However, in a different context and with a different function, the mechanical system could be very interesting to adapt as an autoreactive solution, integrating for instance thermal cylinders in alternative to the pistons.

6.1.5 Strandbeest

The Strandbeest are mechanic animals made out of a system of plastic tubes and connections, animated by the wind as the kinetic energy is transformed into a walking pattern (BBC One 2010). The actuator consists of recycled plastic bottles that collect high pressurized air captured by the wings of the animal and pumped into the system. After a few hours of standing in the wind, the bottles fill up with enough pressurized air for the first walking cycle to start. The "muscles" consist of tubes sliding inside one another, stretching in and out, with a rubber ring on the inner tube acting as an antagonist muscle. These are activated through a domino effect of opening and closing taps, activating one muscle after another. A rod in the shape of a crankshaft, running horizontally to the ground in the middle of each animal, acts as spine converting the rotation movement into the walking. The walking movement varies depending on the mechanical design of the legs, transforming the rod's movement into a walking curve (◘ Fig. 6.5). The legs are composed by 11 rods, the length of which can vary to give different walking

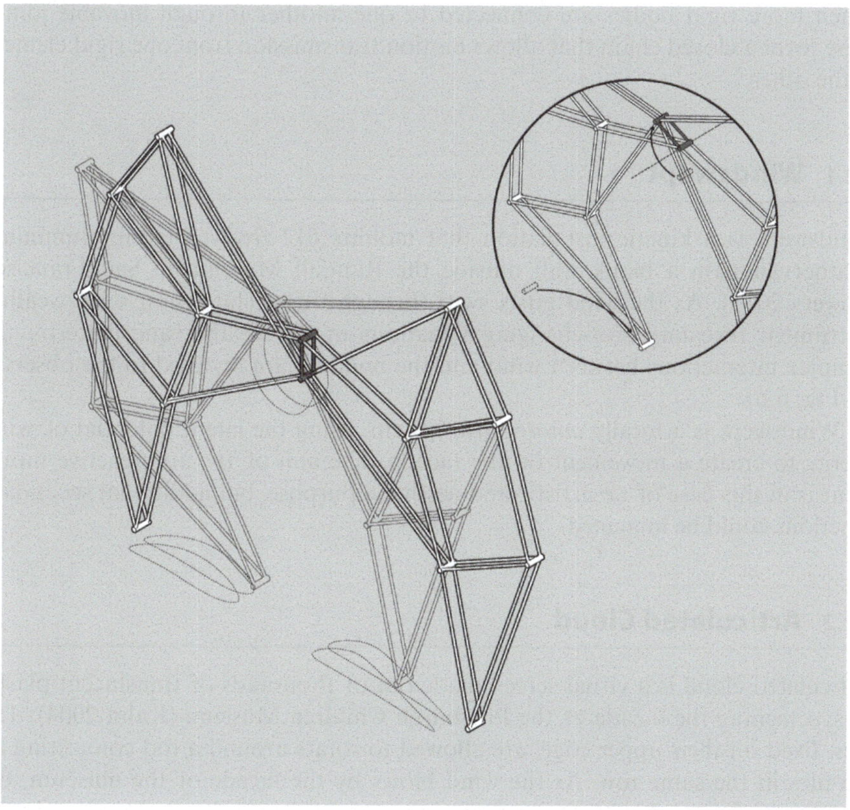

◘ **Fig. 6.5** Scheme of the functioning of a Strandbeest system (Kobas and Koth 2020). Walking curve of a set of legs (left) and detail of the transmission of movement from the central rod to the foot element

curves. The end of the leg structure describes a triangle-shaped movement that support the animal in the stride phase while the other sides of the triangle are responsible for the lifting phase of the leg. The animal is always supported by part of the legs, providing stability, while the others are lifted.

Being powered exclusively by wind energy, the Strandbeest system can be considered to be fully *autoreactive*.

6.2 Rigid Tilting Systems

Kinetic systems using simple tilting movements are quite common being straightforward to control, cheap and very resistant. As the kinetic motion that can be achieved in itself is very elementary, complexity is often achieved by multiplying the number of elements and relying on effects that resemble that of swarm behaviour.

These structures are made with elements that do not alter their geometric proprieties as length, angles or shapes, but are allowed to move with typically one, sometimes two degrees of freedom, mostly using some type of lever system. When more rigid bodies are connected to one another through movable joints, these form a closed chain that allows motion transmission from one rigid element to the other.

6.2.1 Windswept

Windswept is a kinetic installation that mounts 612 freely rotating aluminium weathervanes on a blank wall outside the Randall Museum in San Francisco (Sowers 2011). As the wind gusts swirl through this architectural scale weather instrument revealing ever-changing variations in the weathervane patterns, the complex interactions between wind and the building are revealed to the observer (◻ Fig. 6.6).

Windswept is a totally *autoreactive* system, using the latent potential of wind energy to create a movement on the façade. The aim of the autoreactive movement is in this case of an artistic and aesthetic purpose, but additional secondary functions could be imagined.

6.2.2 Articulated Cloud

Articulated cloud is a visual screen made out of thousands of translucent plastic tiles, screening the façade of the Pittsburgh Children Museum (Kahn 2004). The tiles, fixed on their upper edge, are allowed to rotate around a rod connecting all the tiles in the same row. As the wind blows by the façade of the museum, the tiles move, giving substance to the wind gusts and the turbulences forming ripples on the screen (◻ Fig. 6.7).

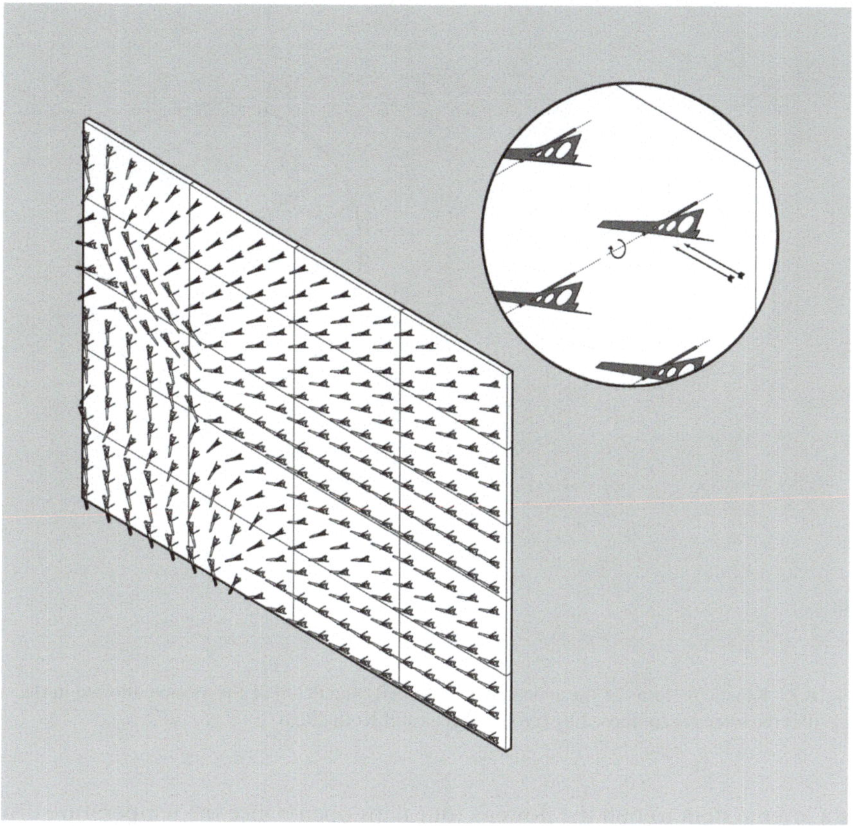

Fig. 6.6 Kinetic patterns of the windswept installation (left) and detail of the rotating weathervanes (Kobas and Koth 2020)

Articulated cloud resembles the windswept installation in many aspects as it is an *autoreactive* artistic installation that is moved by the wind gusts. In this case, however, a second function is added to the aesthetics of the movement: the screening and diffusion of the sunlight on the façade of the museum.

6.2.3 Air Flow(er)

The air flow(er) is a project for an autoreactive ventilating surface made of triangular rigid surfaces arranged into a square flower shape that flips open as the ambient temperature rises, allowing air to pass through (Lift Architects 2007) (Fig. 6.8).

The device is designed to open around 26 °C: the rigid triangular flaps rotate along a fixed edge and are activated by shape-memory alloy (SMA) wires which

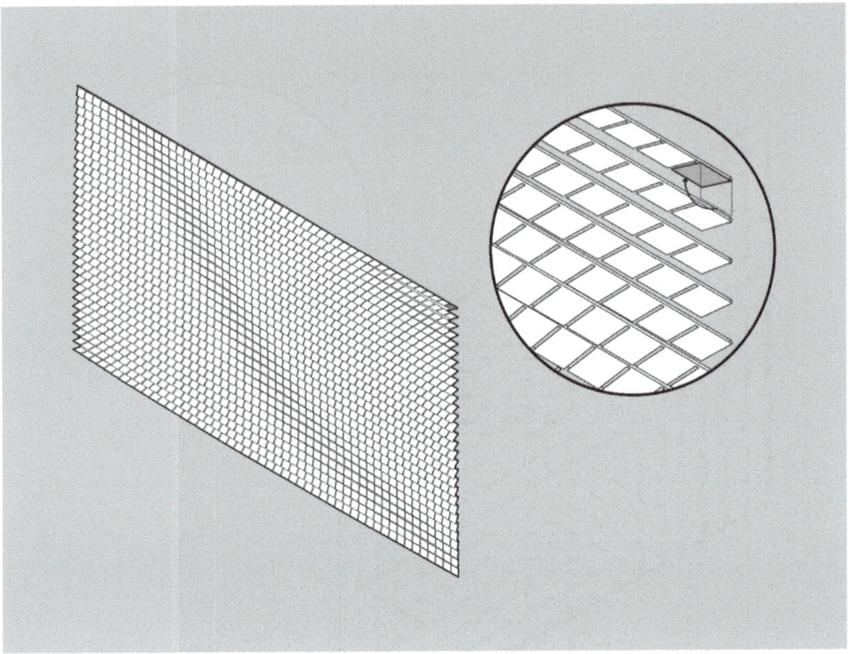

Fig. 6.7 Kinetic patterns of the articulated cloud (left) and detail of the motion allowed in the single tiles the screen is composed by (right) (Kobas and Koth 2020)

use a lever system to pull the flower's four flaps open. Once the temperature drops and the shape-memory wires are deactivated around 15 °C, the flaps are pulled back to their initial position, closing the device, through the action of elastic strings. The system is fully *autoreactive*.

6.2.4 Kinetic Wall

The kinetic wall is an *interactive* installation using 139 kinetic petals that reacts to the presence of users by opening up dynamic windows on the wall (Leva 2018) (Fig. 6.9). The wall is triggered by a detection system that reacts to movements, while kinetic patterns performed are controlled though an animation software.

6.2.5 Pneukox

Pneukox is a master student project for an openable fibre glass wall system. The wall system is subdivided into individually foldable squares that split up into eight rigid triangular surfaces that are folded using a combination of an umbrella-like mechanism and a rotary movement using rigid rods (Keilig et al. 2010) (Fig. 6.10).

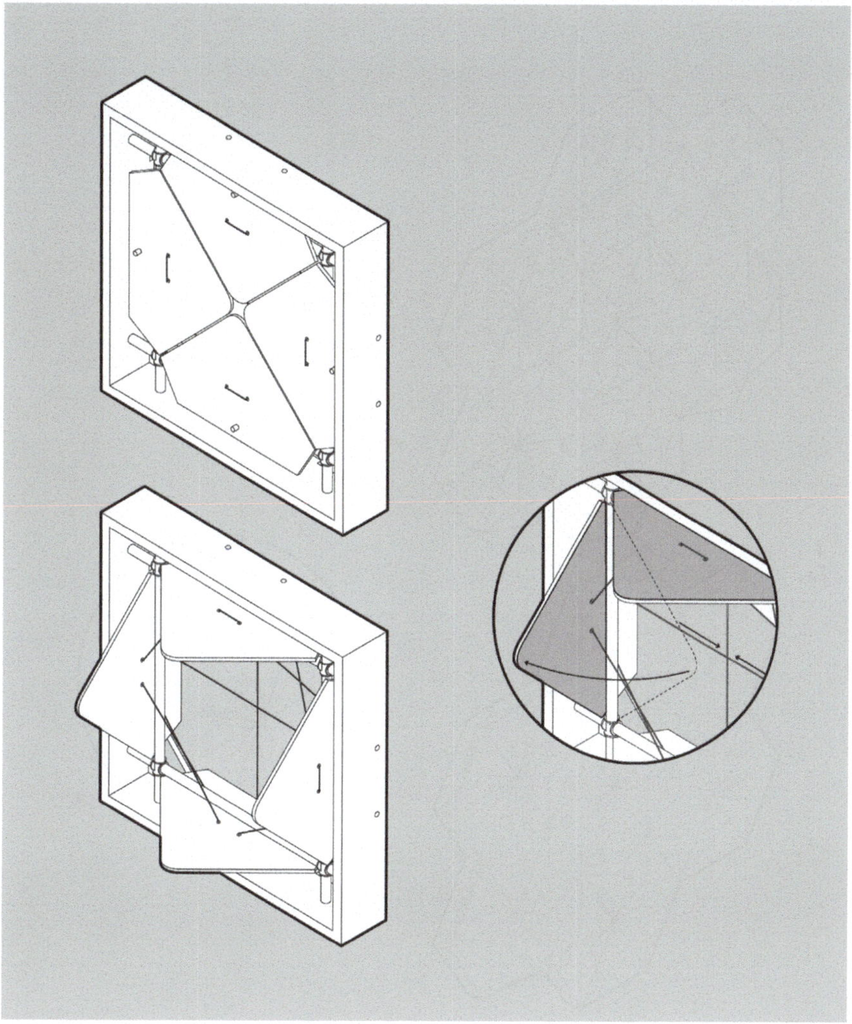

Fig. 6.8 Schemes of the opening kinetics of the air flow(er) system through the operation of wires (left) and detailed view of the opening and closing mechanism (Kobas and Koth 2020)

The system opens up in reaction to changes in the surrounding temperature and is operated using temperature-sensitive waxes that drive a piston mechanism. The shading device is therefore a fully *autoreactive*.

6.2.6 Moucharabieh

The facade of the Arab World Institute (AWI) in Paris realized in the early eighties (IMA 2020) is a high-tech experiment on architectural scale and is largely

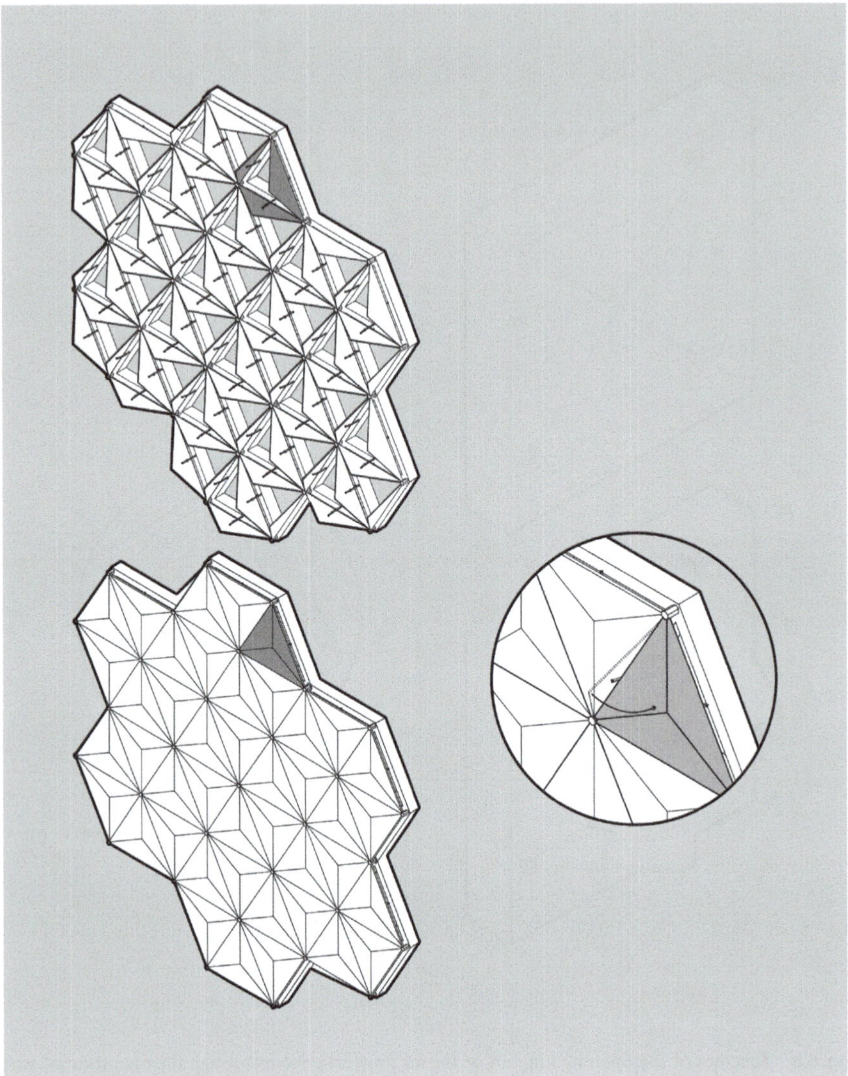

Fig. 6.9 Scheme of the opening and closing of the kinetic wall (left) and detail of the flaps performing the movement (Kobas and Koth 2020)

considered as the forerunner of the family of adaptive facade technologies (Ritter 2007). The double-glass façade integrates a metallic kinetic sunscreen system that adjusts the indoor light patterns depending on the outdoor light conditions at an hourly rate. The kinetic system, which reinterprets the traditional Arabic "Moucharabieh" sunscreens, is realized as a multitude of motor-controlled camera-like metal diaphragms of different sizes (Fig. 6.11).

The diaphragm mechanism drives the shutter surfaces to slide over one another, as in the aperture of a photographic lens. This façade technology is

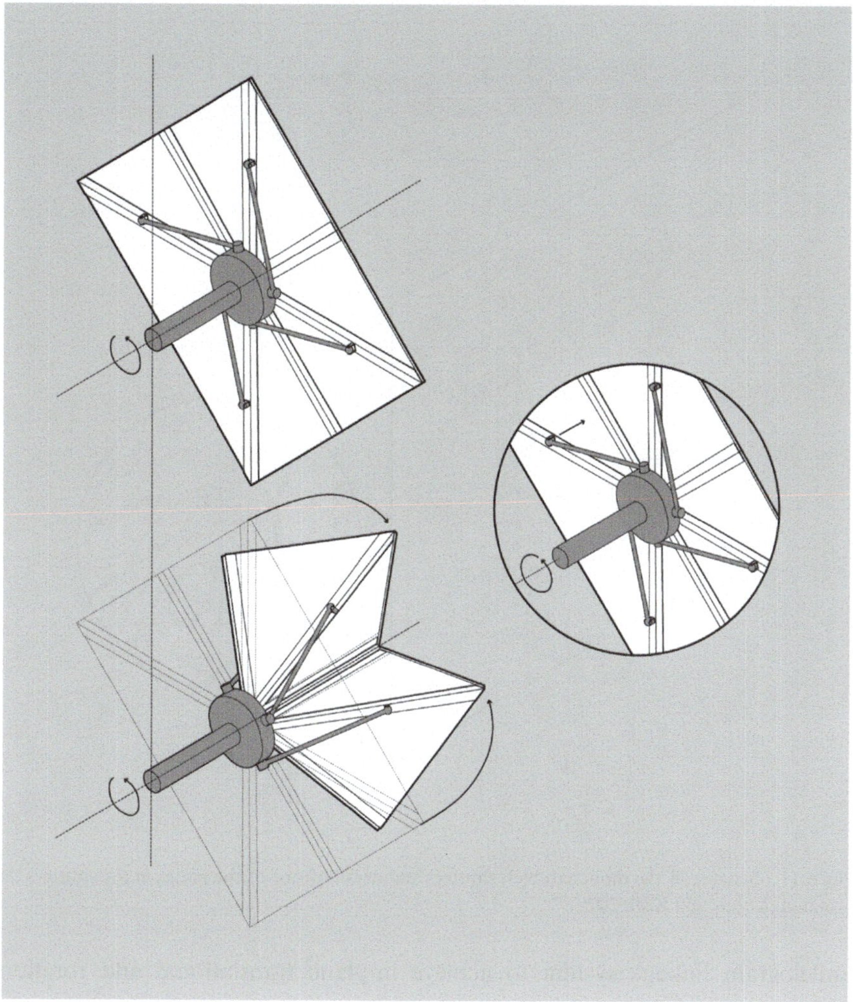

Fig. 6.10 Kinetic operation of the pneukox prototype through the rotation of rigid rods (Kobas and Koth 2020)

equipped with a complex sensor-brain network to process all the incoming information and controlling the single shutters. It is therefore a typical example of a *responsive* system. The simplicity of the mechanical motion transmission through parallel sliding of rigid thin plates remains of interest in an autoreactive context.

6.2.7 Adaptive Fritting

Adaptive Fritting™ and Tessellate™ are two different series of overlapping automated sun shading screens, moved parallelly over each other using a

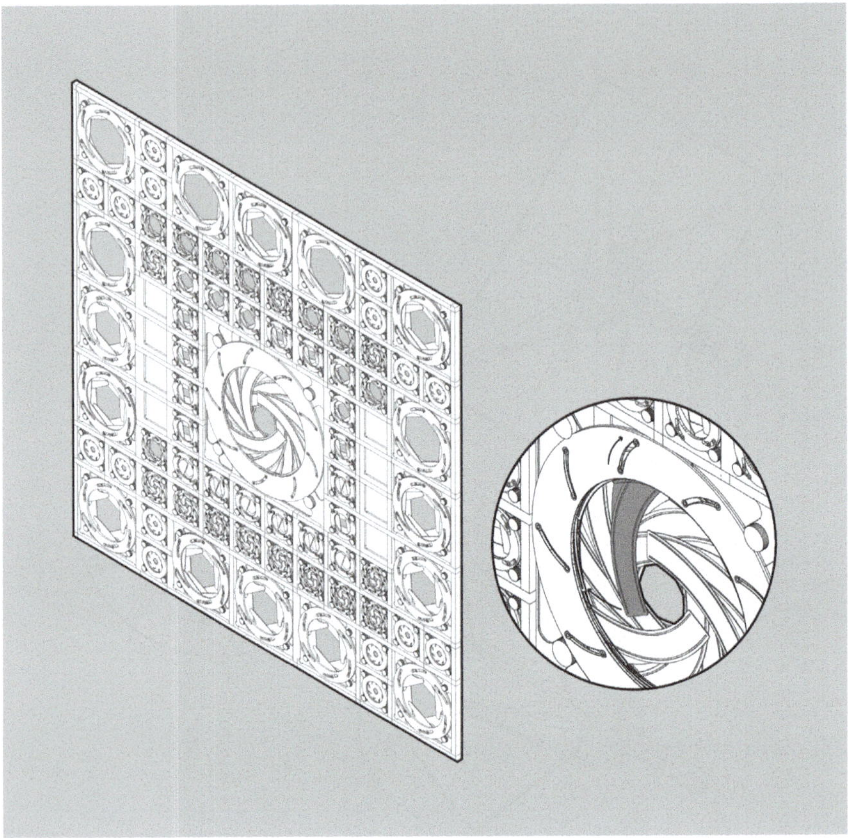

Fig. 6.11 Scheme of the moucharabieh shutters and detail of one of the plates in the shutters that are rotated (Kobas and Koth 2020)

parallelogram linkage system to achieve in-plane translational and rotational motion.

The Tessellate™ screens are parallelly sliding perforated plastic or metal screens integrated in building facades as adjustable shadings which can also regulate ventilation and privacy. This system can be realized with panels in a wide range of two-dimensional and three-dimensional shapes, sizes and patterns.

Adaptive Fritting™ is a kinetic installation realized by Hoberman Associates at the Graduate School of Design at Harvard University (Drozdowski and Gupta 2009). It is made out of a series of printed glasses with dots of different colours that modulate the transparency of the glazed surfaces depending on how the patterns overlap (Fig. 6.12).

Both these kinetic systems are controlled using sensors that convey the environmental information to a simple control centre, which in turn processes the information, determines the reaction in terms of timing and of magnitude and activates the electrical motor that drives the scissor structure to move. The systems are therefore *responsive*.

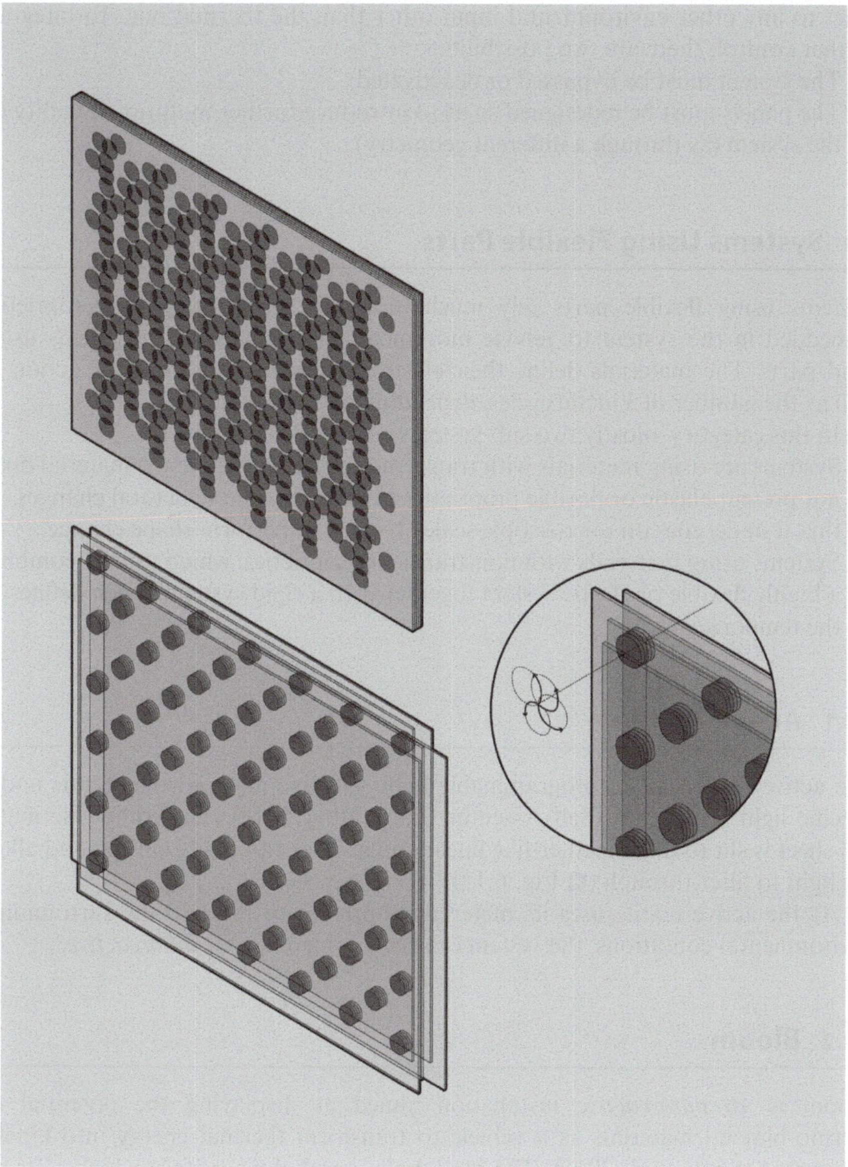

Fig. 6.12 Sliding scheme of the Adaptive Fritting™ system (Kobas and Koth 2020). By moving multiple panels on parallel planes, the adaptive system achieves a progressive variation of transparency or opacity of the surface

The simplicity of the mechanism, however, suggests that these systems have a high potential to be realized following autoreactive principles, using thermally activated materials as a vehicle of the reaction. The main drawback of the autoreactive version of these systems is that it cannot be reprogrammed to selectively react to any other environmental input other than the thermal one. To integrate further control, there are two possibilities:

- The system must be bypassed or deactivated;
- The panels must be redesigned so as to introduce further multifunctionality in the system (as through a different geometry).

6.3 Systems Using Flexible Parts

Systems using flexible parts rely much more on specific material proprieties embedded in the system to vehicle movement, in comparison to systems using rigid parts. The materials define their elastic proprieties, the range of action as well as the number of kinetic cycles, depending on the wear.

In this category, mostly two sub-systems can be identified:

- Systems are using materials with transient proprieties, where the material does not present elastic or flexible proprieties in itself, but the structural changes that it undergoes on microscopic scale are used to perform shape change;
- Systems using materials with non-transient proprieties, which usually combine a highly flexible or elastic system together with a rigid system, which defines the temporary shape.

6.3.1 Active Textile

The active textile uses a programmable light-sensitive material that bends under specific light conditions (Self-Assembly Laboratory 2018). The composite material sheet is slit to form feather-like linear fringes that can easily curl up and allow the light to filter through (◘ Fig. 6.13).

As the active textile uses its material proprieties to react to the surrounding environmental conditions, the system can be considered to be *autoreactive*.

6.3.2 Bloom

Bloom is an *autoreactive* installation aimed at displaying the potential of thermo-bimetal materials as a vehicle to transform thermal energy into kinetic movement (Sung et al. 2011). The hyperbolic paraboloid surface which rests on a frame of folded aluminium profiles is covered with 414 laser cut bimetal panels. These panels are cut into smaller flaps, which open up or close depending on the path of the sun heating up the metallic surfaces (◘ Fig. 6.14).

InVert is a variation of the same system, used as a passive shading system within a curtain wall. The flaps flip into a different position as the temperature rises inside the glazed cavity, preventing the sun rays from entering the building (Sung et al. 2020).

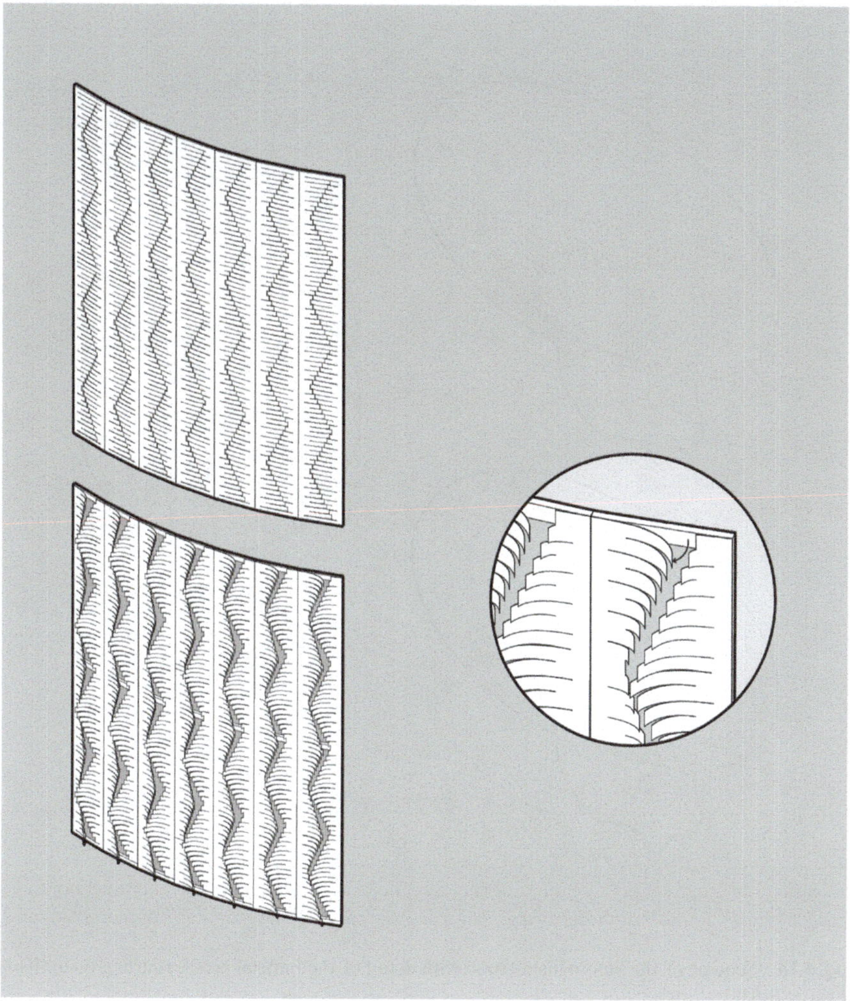

Fig. 6.13 Scheme of the active textile opening up (left) and detail of the curling fringes creating movement on the surface (Kobas and Koth 2020)

6.3.3 Lotus

The lotus dome (Studio Roosegaarde 2011) is an interactive installation made out of hundreds of ultra-light heat sensitive polyester flowers forming a permeable dome structure. As a spectator approaches and sensors detect movement, a light within the dome is lit, causing the flowers to curl open (Fig. 6.15).

As the triggering element for the flowers to curl open is the heat radiated by the light within the dome, which in turn is actuated by sensors that detect the passing observers, it can be concluded that the movement is the outcome of a

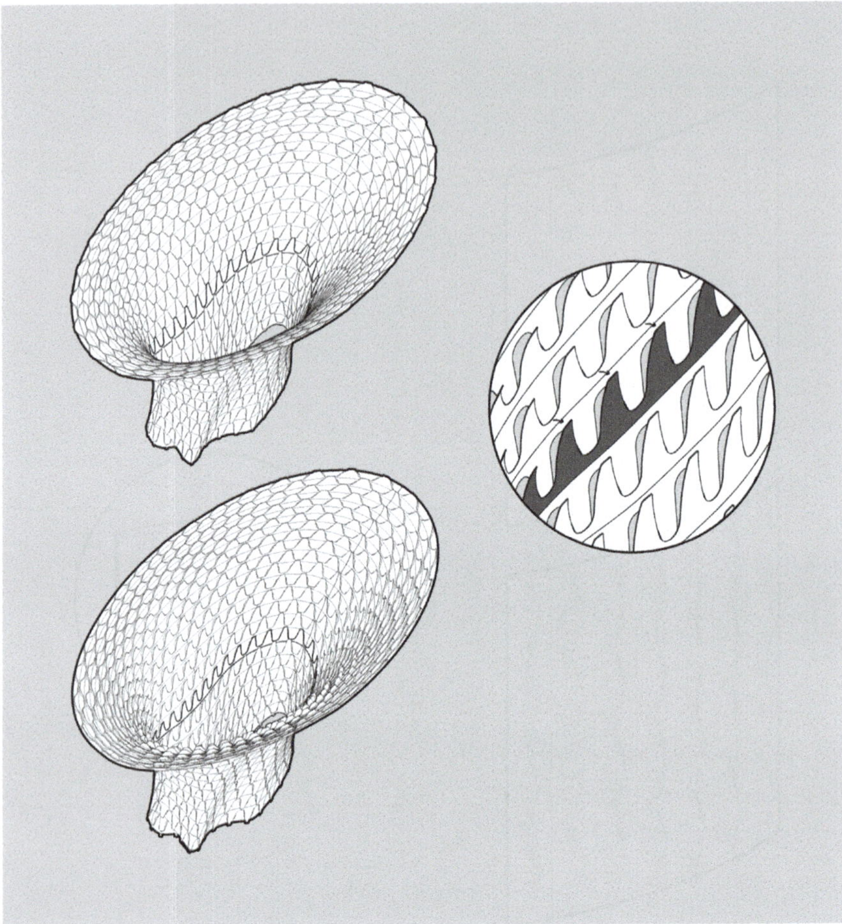

Fig. 6.14 Scheme of the bloom installation with detail of the bimetal profiles curling open (Kobas and Koth 2020)

chain reaction triggered by the sensor-brain system. Therefore, the lotus installation is considered to be an *adaptronic* system.

6.3.4 Hygroscope

The hygroscope installation is very similar in its geometry and kinetic pattern to the Lotus installation. The main difference is that Hygroscope explores the potential of a hygroscopic (humidity-sensitive) material to achieve kinetic response and is therefore a fully *autoreactive* system (Fig. 6.16).

The kinetic parts are thin triangular surfaces made out of a maple veneer and synthetic composite, which is programmed to react differently depending on the fibre direction, the length, thickness and geometry (ICD 2002). The model is sus-

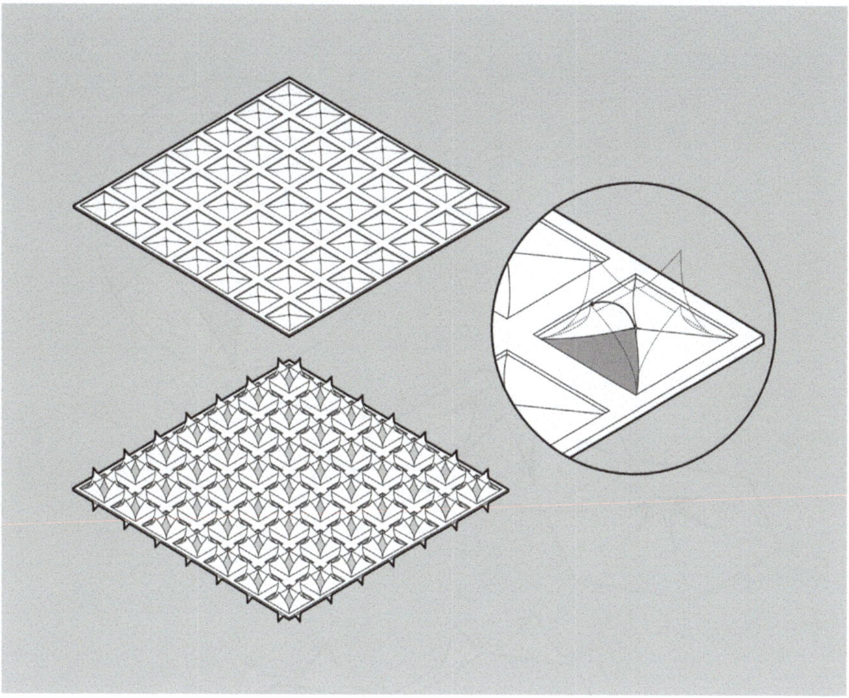

Fig. 6.15 Scheme of the lotus flowers' curling represented, not as a dome structure but as a planar surface, with the detail of the flaps curling open (Kobas and Koth 2020)

pended in a showcase where the RH is regulated through an accelerated database of the relative humidity in Paris. Absorption of humidity causes the distance between the fibres of the wood to increase, resulting in a swelling or lengthening of the material in the direction of the fibres.

6.3.5 Homeostatic

The homeostatic façade system is made of thin stripes integrating an electroactive polymer (EAP) wrapped around a flexible core, allowing the section of the stripes to bend with the expansion or contraction of the polymer (Decker 2013) (Fig. 6.17).

The plates are coated with electrodes on two sides and a silvered layer (for distributing the electrical charge on the whole surface), controlling the deformation of the polymer core. The actuation of one surface of the plates creates a bending and a deformation effect similar to that of a bi-metal, which is used to progressively shade the facade for light, solar and heat control.

The movement in this system being directly controlled through a sensor-brain system integrating a functional material, and the system can be considered to be an *adaptronic* system. The use of a bi-layered composite system driving kinetic

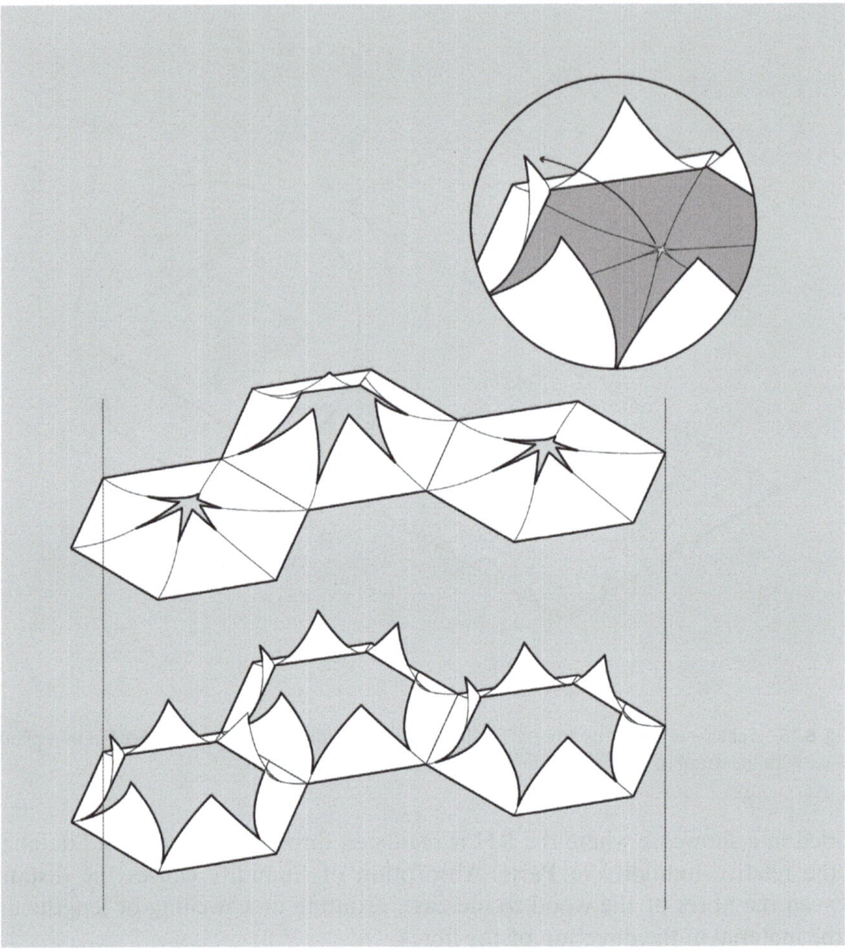

Fig. 6.16 Scheme of the curling veneer flaps in the hygroscope system (Kobas and Koth 2020)

change through the extension or shrinking of one of the layers is, however, a simple mechanism that is employed in many autoreactive systems.

6.3.6 Flectofin

Flectofin® is a hinge-less sun shading device that was the outcome of a bionic study of the *Strelitzia reginae* flower's elastic and structural geometric deformation upon application of a weight in specific points. The system has a linear geometry with two thin shell elements attached orthogonally to a glass fibre reinforced polymer (GFRP) beam element (Lienhard et al. 2011). The uniaxial bending of the beam causes an asymmetrical bending and torsional buckling of the shell elements which flap open (Fig. 6.18).

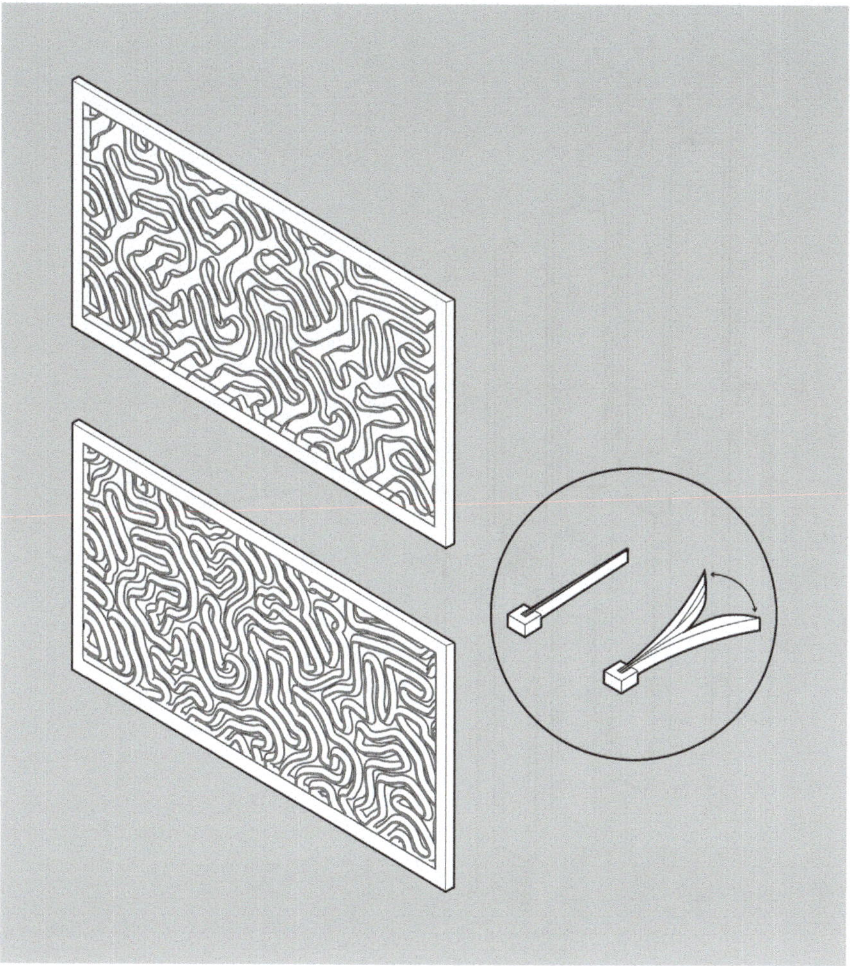

Fig. 6.17 Scheme of the homeostatic façade kinetic transformation (left) and sectioned detail of the strips achieving the kinetic change (right) (Kobas and Koth 2020)

The actuation of the beam element is achieved by actively controlling the operation of a motorized device. The overall system is therefore considered to be a *responsive* system.

6.3.7 Kiemenbox

The Kiemenbox is a master student project where a wooden structure has been covered with a flexible polyethylene fabric, with cleaved openings shaped as fish gills (hence "*Kiemen*"—gill in German) (Schmidt and Schneider 2013). The edges of the openings are equipped with a bi-material polymer that bends under specific

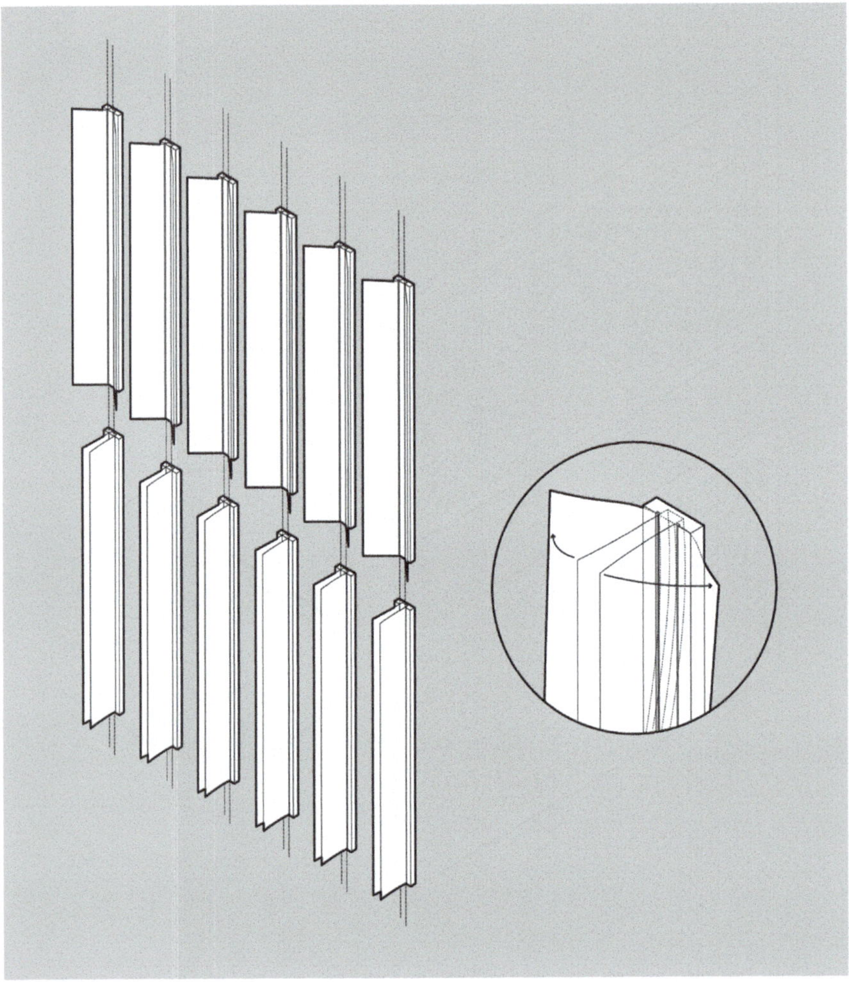

Fig. 6.18 Kinetic diagram of the opening of the shell elements in the Flectofin system (Kobas and Koth 2020)

thermal conditions. As the change in the surrounding air temperature rises, the gills bend, opening up the surface of the box, allowing light to filter through and air to pass (Fig. 6.19).

The system is powered exclusively though the temperature changes in the surrounding air and is therefore fully *autoreactive*.

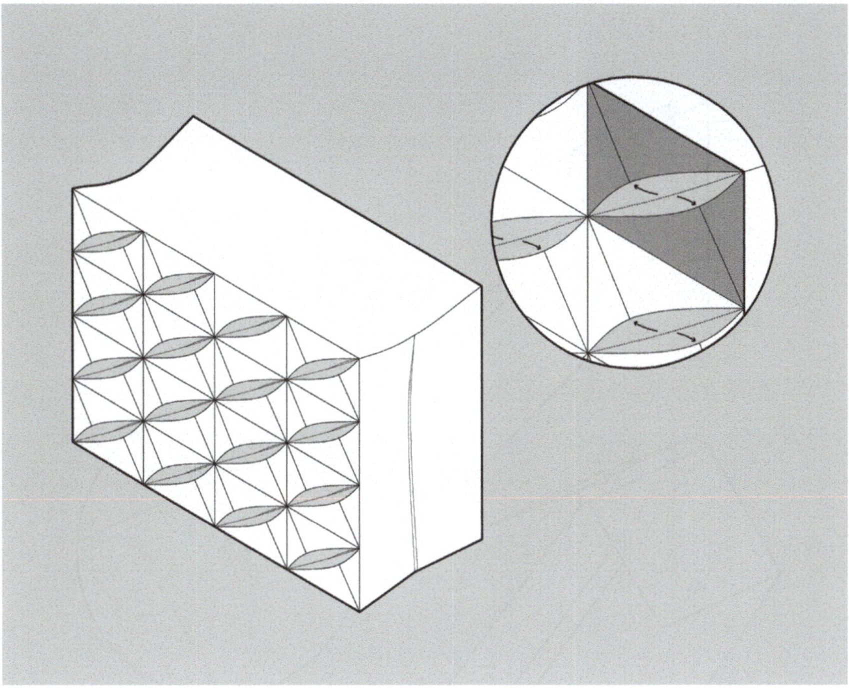

Fig. 6.19 Scheme of the Kiemenbox (left) and kinetic change in the flexible gills (right) (Kobas and Koth 2020)

6.4 Surface Structures

Surface structures are in many cases made out of rigid or flexible smaller units connected to each other. Movement in one shape drives the movement in the next one, with the main advantage that bigger areas can be activated by intervening only punctually with a kinetic input, optimizing the potential output of the system. Among the most common drawbacks of these systems is however the added weight due to the larger surfaces in motion, often requiring more powerful actuating systems.

6.4.1 Shade(w)ing

Shade(w)ing is a master student project of a foldable sunscreen element, inspired of the wing folding in insects, and is shaped using the kinetic proprieties of the basic Miura-ori folding sequence (Persiani et al. 2016) (Fig. 6.20).

The prototype is realized as a single thin composite surface: two aluminium sheets enclose a highly elastic material which creates flexible hinges when exposed through milling, allowing the sheet to be bent along the folding lines. The shading device unfolds the activated by a thermal cylinder and is therefore a fully *autoreactive*.

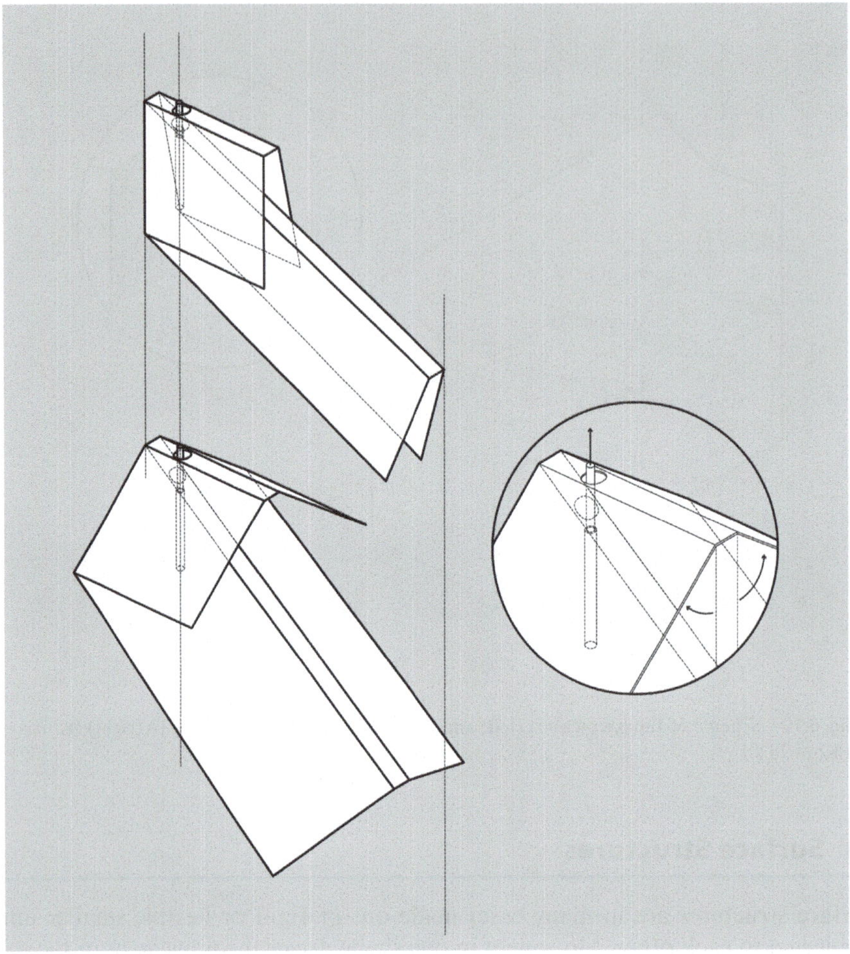

Fig. 6.20 Kinetic operation of the Shade(w)ing prototype using a thermal cylinder (Kobas and Koth 2020)

6.4.2 Moving Mesh

The moving mesh is a master student project for an adaptive sun shading realized with the same flexible aluminium composite as Shade(w)ing. The system works as a single perforated surface where the folded geometry drives the opening and closure of diamond-shaped flaps (Salvi and Huygels 2016) (Fig. 6.21).

The system is moved using sets of thermal cylinders and is therefore considered to be *autoreactive*, although the actuation itself was not realized in the prototype.

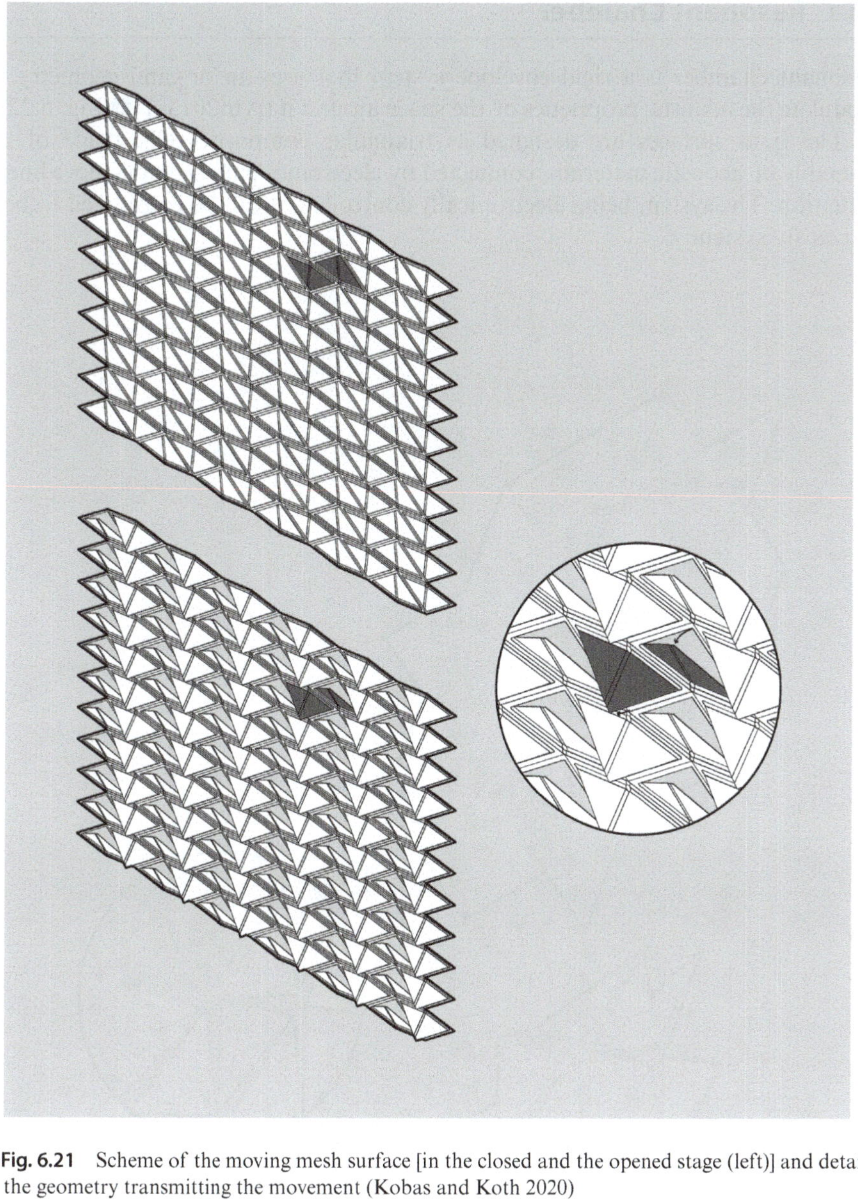

Fig. 6.21 Scheme of the moving mesh surface [in the closed and the opened stage (left)] and detail of the geometry transmitting the movement (Kobas and Koth 2020)

6.4.3 Resonant Chamber

Resonant chamber is a rigid envelope system that uses an origami geometry to modulate the acoustic proprieties of the space around it (rvtr 2013a) (◘ Fig. 6.22).

The rigid surfaces are designed as triangular composite cells, made of an assembly of acoustic materials, connected by electronics panels that achieve linear actuation. The system being electronically controlled, it can be considered to be a *responsive* system.

◘ **Fig. 6.22** Unfolding movements of the resonant chamber surface geometry (Kobas and Koth 2020)

6.4.4 Self-adaptive Membrane

The self-adaptive membrane uses a similar kinetic principle as the surface of the resonant chamber, although the origami folding is geometrically different, and the system is fully *autoreactively* operated (Fig. 6.23).

To deploy the surface, the joints are operated through the action of Nitinol (NiTi) wires, a shape-memory alloy (Gonzalez and More 2015). The wires run through the flat squares of the membrane, subdivided into clusters of four squares which are actuated together. The overall surface changes shape progressively, expanding uniformly in two dimensions as the wires are actuated simultaneously.

6.4.5 Infundibuliforms

Infundibuliforms is a 30 m^2 prototype lightweight kinetic tensile surface. The cable net geometry is extruded from a thermoplastic elastomer on a flat surface and is then stretched into anticlastic forms through the use of outer ovoid and inner weighted rings (rvtr 2016) (Fig. 6.24).

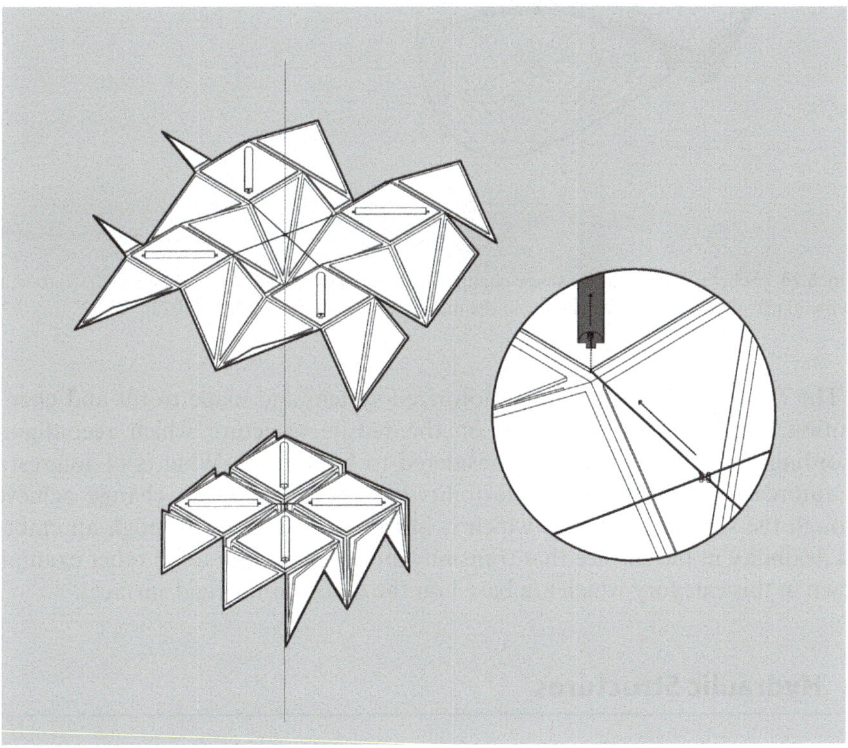

Fig. 6.23 Geometric change of the actuated self-adaptive membrane (left) and detail of the integrated NiTi wires (Kobas and Koth 2020)

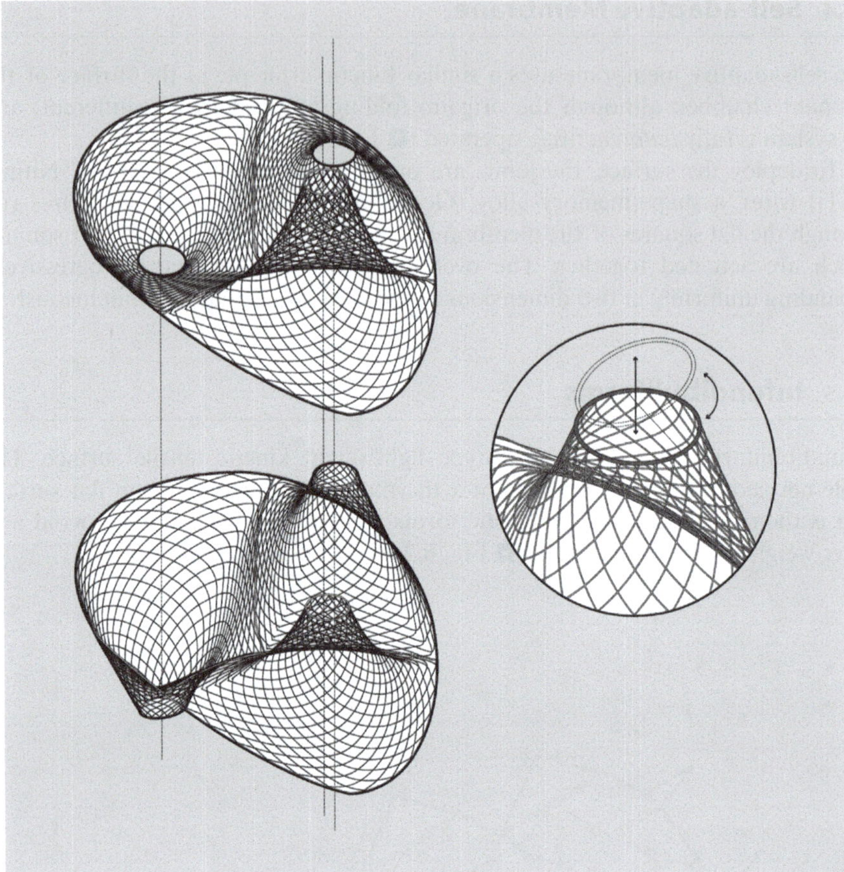

Fig. 6.24 Scheme of the kinetic shape change of the Infundibuliforms installation (left) and detail of a metal ring driving the shape change in the mesh (right) (Kobas and Koth 2020)

The rings are controlled by a motorized system and made to tilt and change position, changing the geometry of the tensile structure which reconfigures accordingly. The system can be considered to be *reactive*. What is of interest in an autoreactive context is the flexibility and potential shape change achieved through the surface geometry, which is in this case achieved through an embedded flexibility in the surface that transmits motion, contrary to all other examples shown in this category which are based on the dynamics of rigid surfaces.

6.5 Hydraulic Structures

The category of hydraulic systems has been found to be the least explored so far, as examples are relatively few. There might be many reasons for this.

One reason could be that the use of systems using rigid or flexible parts to move might simply be more obvious to us: those are the type of movements we are the most familiar with as we see them in many systems that thrive in our environment on land. In the same way as we tend to know better and sympathize with organisms that resemble us in our own vertebrate biomechanics, our first experiments building a kinetic mechanism will rather be based on our own body structure than on the mechanics of a jellyfish that are not particularly well known to us.

Another reason might be that the technical complexity of building an operating pneumatic structure might be overwhelming in the experimental and design phases. Building a mockup of a hydraulic structure is tendentially time and resource consuming as it requires an airtight envelope as well as the shaping of the desired geometry. This implies that the geometrical design process of a hydraulic structure happens, in comparison with other structural typologies, mainly on an intellectual level, reducing the building of mockups to a minimum.

A third challenge is the control and management of the movements in a hydraulic structure. This aspect is clearly a very strong driver in the design: as all the examples discussed in the following part demonstrate, the definition of the rigid parts in the design of the structure is central to control the dynamic patterns of the structure.

In general terms, once these technical and knowledge gaps are filled, the design of hydraulic systems enables many interesting and potentially smart applications with quite varied functions, ranging from variable aperture control, air exchange and thermal control.

6.5.1 Cyclebowl

The façade of the Experiment Cyclebowl Pavilion at Expo Hannover 2000 is made out of a series of three-layered polymeric foil cushions, where the middle cushion is movable by changing the pneumatic pressure between the layers (Brückner 2001). A positive and a negative leaf pattern is printed on the outer two layers of the cushions, allowing to regulate the translucency of the façade by controlling the overlapping of the printed foils (◘ Fig. 6.25).

The operation of the cushions is integrated with the technical system of the building itself. The system can therefore be considered to be *responsive*. What is of interest is however the potential big impact (shading and energy optimization) that can be achieved by integrating such a very simple kinetic movement within a building system. Moreover, such a simple kinetic mechanism has the potential to be transferred within an autoreactive system: for instance, integrating the cushions with shape-memory materials or actuating them through thermal cylinders.

6.5.2 Nervous Ether

The Nervous Ether is a full-scale cellular pneumatic prototype developed within a series of experiments ("Pneusystems") that explore the potential of geometrical and functional operation principles of cellular pneumatic systems (rvtr 2013b).

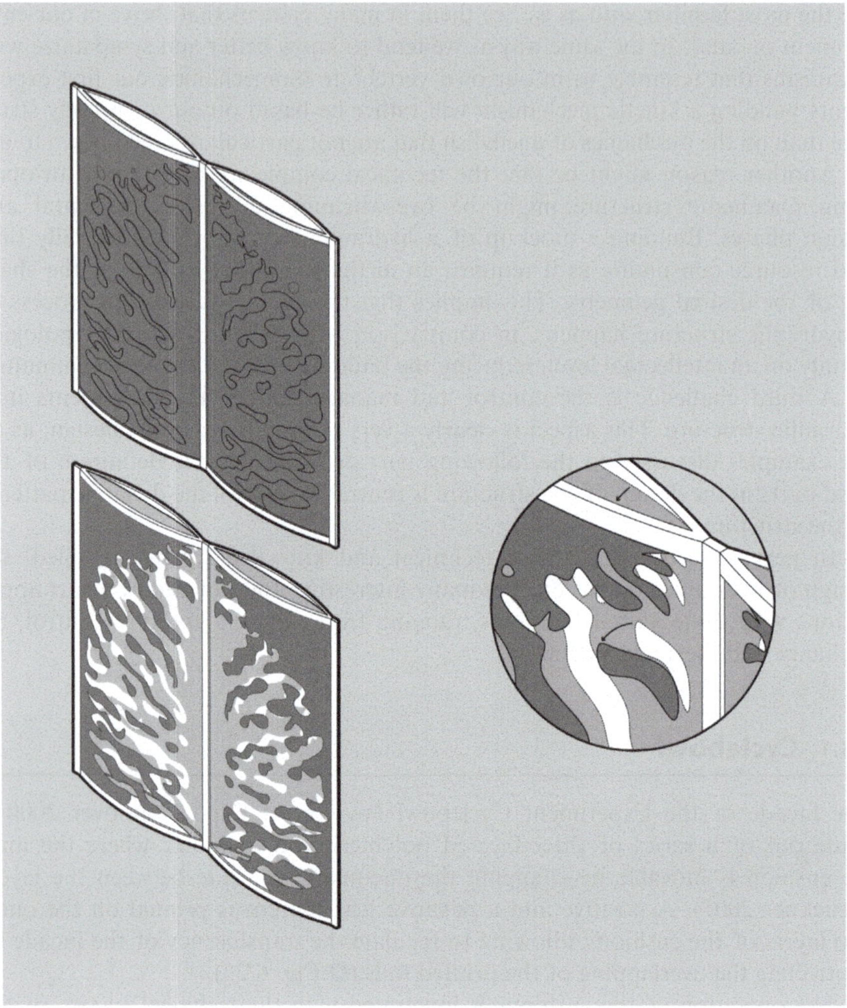

Fig. 6.25 Schematic detail of the pneumatic cushions composing the façade of Experiment Cyclebowl with the different translucent effects achieved by modulating the superposition of the patterns (left) and detail of the interior printed surface that is moved (right) (Kobas and Koth 2020)

Nervous Ether is designed as a tessellated array of intertwined tetrahedral polyethylene cushions, inflated to a constant air pressure, which allow the shapes to interconnect and form a tensioned membrane structure (rvtr 2013c) (Fig. 6.26).

Change in the cushions' shape, density and structural proprieties is achieved using a sensor network to control the internal air pressure of the cushions. The design is therefore *responsive*.

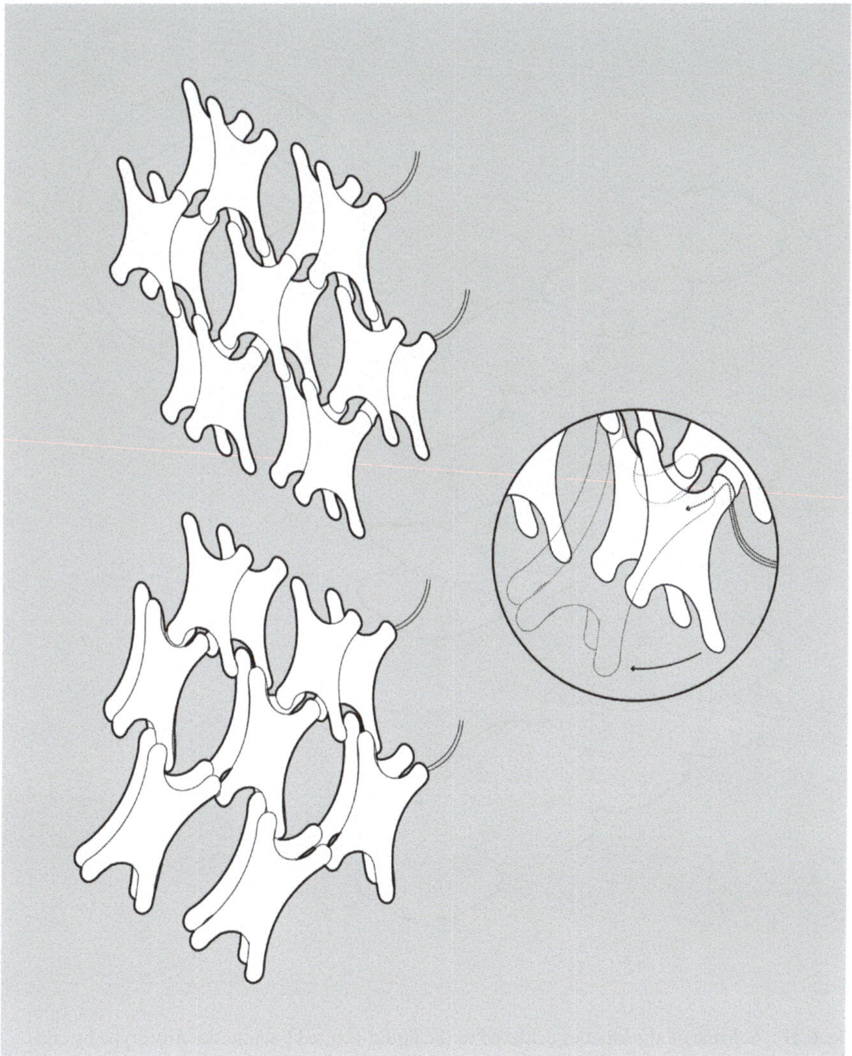

■ **Fig. 6.26** Scheme of the kinetic change achieved in the Nervous Ether prototype by controlling the interior air pressure (Kobas and Koth 2020)

6.5.3 Liquid Printed Pneumatics

Liquid-printed pneumatics is a system made of a series of inflatable silicone cushions, internally connected to each other as in an ice-cube bag (Self-Assembly Laboratory and BMW Design 2018). By controlling the air pressure within each bag, the shape, function and stiffness of the system can be modelled (■ Fig. 6.27).

The operation of the air-pressure control that is at the base of the kinetics in the system is not clear. Assuming that there is a separate air supply system to control the geometry, the case study is considered to be a *responsive* system.

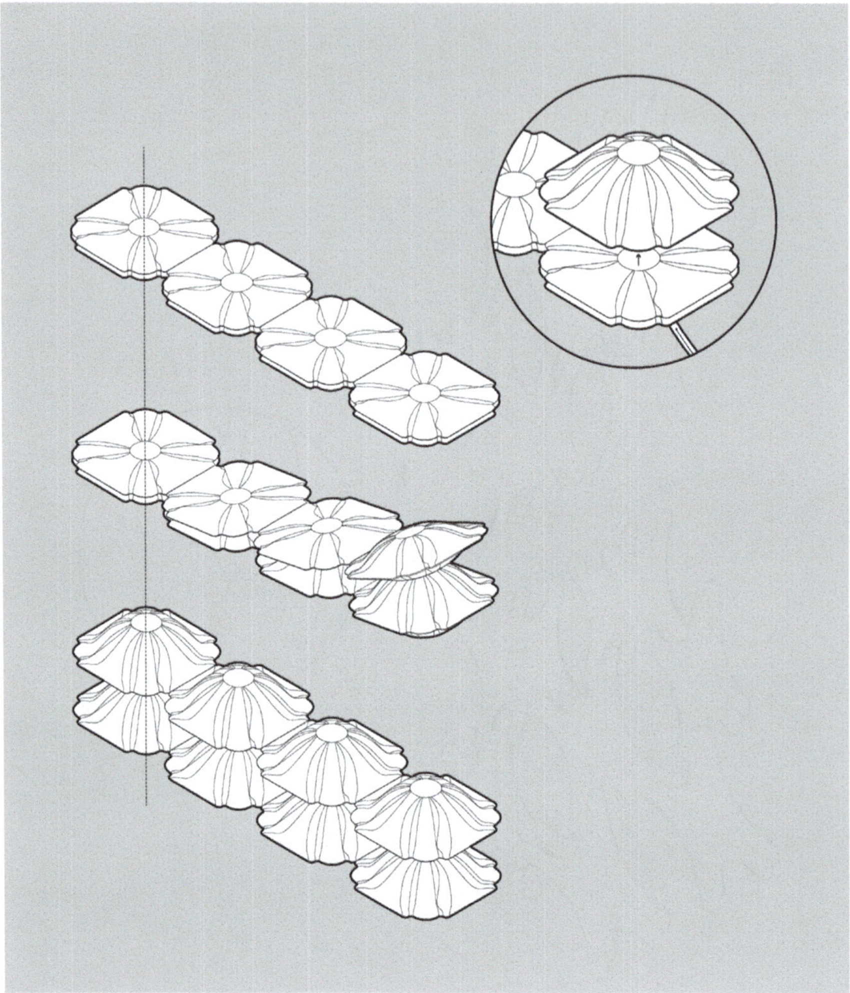

Fig. 6.27 Scheme of the kinetics achieved in the liquid-printed pneumatics prototype by controlling the interior air pressure (Kobas and Koth 2020)

References

Able C, Burson A, Chan I, Chang J, Harris A, Hollyn-Taub T, Lee B, Montgomery E, Roether V, Zhai D (2011) Polymorphic bench. Graduate School of Architecture, Planning and Preservation (GSAPP), Columbia University. New York, USA. Retrieved on 24 Apr 2020 from ► https://vimeo.com/23925803

BBC One (2010) Theo Jansen's Strandbeests. Wallace & Gromit's world of invention episode 1 preview—BBC one, Youtube. Retrieved on 10 Feb 2015 from ► http://www.youtube.com/watch?v=HSKyHmjyrkA

Brückner A (2001) Experiment cyclebowl, av edn. GmBH, Ludwigsburg. ISBN: 3-92-9638-65-7

Decker M (2013) Emergent futures: nanotechology and emergent materials in architecture. In: Zarzycki A, Dermody RJ (eds) Tectonics of teaching. Proceedings of the 2013 building technology educators' society (BTES) conference. Roger Williams University, Bristol, RI, 11–13 July 2013, pp 217–223. ISBN 978-0-9895980-0-2

Drozdowski Z, Gupta S (2009) Adaptive fritting as case exploration for adaptivity in architecture. In: Proceedings of the 29th annual conference of the association for computer aided design in architecture (ACADIA), pp 105–109

Fox MA, Kemp M (2009) Interactive architecture. Princeton Architectural Press, New York

Fu F (2018) Design and analysis of tall and complex structures. Butterworth-Heinemann, Elsevier Ltd. ► https://doi.org/10.1016/C2015-0-06071-3

Gilbert M (2018) Solstice, the kinetic clock. Animaro. Retrieved on 24 Apr 2020 from ► http://www.animarodesign.com

Gonzalez N, More S (2015) Self-adaptive membrane, a passive kinetic system. RS3. Digital matter. Institute for Advanced Architecture of Catalonia (IaaC), Barcelona, Spain. Retrieved on 25 Apr 2020 from ► http://www.iaacblog.com/programs/self-adaptive-membrane-_-a-passive-kinetic-system/

Heatherwick T (2002) Rolling bridge, Heatherwick studio. Paddington Basin, London, UK. Retrieved on 24 Apr 2020 from ► http://www.heatherwick.com/project/rolling-bridge/

ICD (2002) Hygroscope metereosensitive morphology, ICD research buildings/prototypes. ICD University of Stuttgart. Retrieved on 5 Feb 2019 from ► https://www.icd.uni-stuttgart.de/projects/hygroscope-meteorosensitive-morphology/

IMA (2020) Les façades Nord et Sud, Le projet, Architecture. Institut du Monde Arabe. Retrieved on 24 Apr 2020 from ► https://www.imarabe.org/fr/architecture

Jun JW, Alcover J, Silverio M (2015) Remembrane. Digital matters/intelligent constructions. Institute for Advanced Architecture of Catalonia (IAAC). Retrieved on 24 Apr 2020 from ► http://www.iaacblog.com/projects/remembrane-ji-won-jun-josep-alcover-matteo-silverio/, and ► https://issuu.com/remembrane/docs/remembrane_book_publish

Kahn N (2004) Articulated cloud, Pittsburgh children's museum. Ned Kahn Studios, Pittsburgh, PA. Retrieved on 5 Apr 2020 from ► http://nedkahn.com/portfolio/articulated-cloud

Keilig A, Franke R, Steinbüchel A, Arnold M (2010) Pneukox. Kunststoffobjekt, htwk Leipzig. Retrieved on 25 Apr 2020 from ► https://kunststoffobjekt.com/2013/04/27/pneukox/

Kobas B Koth SC (2020) Autoreactive case studies. Unpublished images

Kretzer M, Hovestadt L (eds) (2014) ALIVE: advancements in adaptive architecture, vol 8. Applied virtuality book series. Birkhauser Verlag GmbH, Basel. ISBN: 978-3-99043-667-7

Leva (2018) Kinetic wall. Leva and TODO, Cupra's booth. In: 88th Geneva international motor show. Retrieved on 29 Apr 2020 from ► https://www.leva.io/projects/kinetic-wall

Lienhard J, Schleicher S, Poppinga S, Masselter T, Milwich M, Speck T, Knippers J (2011) Flectofin: a hingeless flapping mechanism inspired by nature. Bioinspir Biomim 6(4):045001. ► https://doi.org/10.1088/1748-3182/6/4/045001

Lift Architects (2007) The air flow(er), lift architects. Retrieved on 10 Apr 2016 from ► http://www.liftarchitects.com/air-flower

Oosterhuis K (2012) Hyperbody forward to basics. Hyperbody 2012. Retrieved on 11 Nov 2014 from ► http://www.hyperbody.nl/fileadmin/pdf/HyperbodyMasterBooklet_2012.pdf

Persiani SGL (2018) Biomimetics of motion. Nature-inspired parameters and schemes for kinetic design. Springer International Publishing, Cham, Switzerland. ► https://doi.org/10.1007/978-3-319-93079-4

Persiani SGL, Battisti A (2018) Frontiers of adaptive design, synthetic biology and growing skins for ephemeral hybrid structures, In: Membrane technology—energy efficiency and sustainability. InTechOpen. ► https://doi.org/10.5772/intechopen.80867

Persiani SGL, Molter PL, Aresta C, Klein T (2016) Mapping of environmental interaction and adaptive materials for the autoreactive potential of building skins. In: Proceedings of the 41st IAHS world congress sustainability and innovation for the future. Algarve, Portugal, 13–16 Sept 2016, p 302. ISBN 978-989-98949-4-5

Ritter A (2007) Smart materials in architecture, interior architecture and design. Birkhaüser, Basel. ISBN 13: 978-3-7643-7327-6 rvtr (2013a) Resonant chamber, material system prototype (acoustics). rvtr, Taubman College of Architecture and Urban Planning, University of Michigan. Retrieved on 25 Apr 2020 from ► http://www.rvtr.com/projects/resonant-chamber

rvtr (2013b) Pneusystems, material system prototype (air). rvtr, Taubman College of Architecture and Urban Planning, University of Michigan. Retrieved on 25 Apr 2020 from ► http://www.rvtr.com/projects/pneusystems

rvtr (2013c) Nervous ether, inflatables workshop: prototype development and exhibition. rvtr, Taubman College of Architecture and Urban Planning, University of Michigan. Retrieved on 25 Apr 2020 from ► http://www.rvtr.com/projects/nervous-ether

rvtr (2016) Infundibuliforms, responsive envelopes: kinetic tensile surface environments. rvtr, Taubman College of Architecture and Urban Planning, University of Michigan. Retrieved on 25 Apr 2020 from ► http://www.rvtr.com/projects/infundibuliforms

Salvi R, Huygels B (2016) The moving mesh. Performative Folding Master Course, Associate Professorship of Architectural Design and Building Envelope, Technical University of Munich. In: Persiani SGL, Battisti A (2018) Frontiers of adaptive design, synthetic biology and growing skins for ephemeral hybrid structures. In: Energy-efficient approaches in industrial applications. IntechOpen. ► https://doi.org/10.5772/intechopen.80867

Schmidt S, Schneider C (2013) Kiemenbox. Kunststoffobjekt im Handgepäck, htwk Leipzig. Retrieved on 25 Apr 2020 from ► https://kunststoffobjekt.com/2013/02/01/140/

Self-Assembly Lab, BMW Design (2018) Liquid printed pneumatics, the future starts here, sainsbury gallery at the V&A Museum, London. MIT's Self-Assembly Lab. Retrieved on 25 Apr 2020 from ► https://selfassemblylab.mit.edu/active-textile

Self-Assembly Lab, Steelcase, Designtex (2018) Active textile, Cooper Hewitt, Smithsonian design museum—the senses: design beyond vision exhibition. MIT's Self-Assembly Lab. Retrieved on 25 Apr 2020 from ► https://selfassemblylab.mit.edu/active-textile

Sowers C (2011) Windswept, public art and museum installations. Retrieved on 24 Apr 2020 from ► https://www.charlessowers.com/new-page-1

Studio Roosegaarde (2011) Lotus Dome, Studio Roosegaarde. Sainte Marie Madeleine Church, Lille, France. Retrieved on 24 Apr 2020 from ► https://www.studioroosegaarde.net/project/lotus-dome

Sung D, Ramirez A, Phillips L, Michalski J, Xu W, Kang J, East M, Ho E (2020) InVert auto-shading windows, DOSU studio architects. M&A Gallery, Silver Lake, CA. Retrieved on 24 Apr 2020 from ► https://www.dosu-arch.com/invert

Sung D, Wood D, Butterworth K, Chen A, Ganis R, Greene D, Michalski J, Morinaga S, Shieh E (2011) Bloom, DOSU studio architects. M&A gallery. Silver Lake, CA. Retrieved on 24 Apr 2020 from ► https://www.dosu-arch.com/bloom

The manufacturer's authorised representative in the EU is Springer Nature Customer Service Centre GmbH, Europaplatz 3, 69115 Heidelberg, Germany. If you have any concerns regarding our products, please contact ProductSafety@springernature.com

Printed and bound by CPI Group (UK) Ltd, Croydon, CR0 4YY
30/09/2026
02225559-0004